ENVIRONMENTAL SCIENCE, ENGINEERINGAND TECHNOLOGY SERIES

DIRECTORY OF CONSERVATION FUNDING SOURCES FOR DEVELOPING COUNTRIES: CONSERVATION BIOLOGY, EDUCATION AND TRAINING, FELLOWSHIPS AND SCHOLARSHIPS

ENVIRONMENTAL SCIENCE, ENGINEERING AND TECHNOLOGY SERIES

The Amazon Gold Rush and Environmental Mercury Contamination
Daniel Marcos Bonotto and Ene Glória da Silveira
2009. ISBN: 978-1-60741-609-8

Environmental Effects of Off-Highway Vehicles
Douglas S. Ouren, Christopher Haas, Cynthia P. Melcher, Susan C. Stewart, Phadrea D. Ponds, Natalie R. Sexton, Lucy Burris, Tammy Fancher and Zachary H. Bowen
2009. ISBN: 978-1-60692-936-0

Nitrous Oxide Emissions Research Progress
Adam I. Sheldon and Edward P. Barnhart (Editors)
2009. ISBN: 978-1-60692-267-5

Agricultural Runoff, Coastal Engineering and Flooding
Christopher A. Hudspeth and Timothy E. Reeve (Editors)
2009. ISBN: 978-1-60741-097-3

Fundamentals and Applications of Biosorption Isotherms, Kinetics and Thermodynamics
Yu Liu and Jianlong Wang (Editors)
2009. ISBN: 978-1-60741-169-7

Conservation of Natural Resources
Nikolas J. Kudrow (Editor)
2009. ISBN: 978-1-60741-178-9

Forest Canopies: Forest Production, Ecosystem Health and Climate Conditions
Jason D. Creighton and Paul J. Roney (Editor)
2009. ISBN: 978-1-60741-457-5

Directory of Conservation Funding Sources for Developing Countries: Conservation Biology, Education and Training, Fellowships and Scholarships
Alfred O. Owino and Joseph O. Oyugi
2009 ISBN: 978-1-60741-367-7

ENVIRONMENTAL SCIENCE, ENGINEERINGAND TECHNOLOGY SERIES

DIRECTORY OF CONSERVATION FUNDING SOURCES FOR DEVELOPING COUNTRIES: CONSERVATION BIOLOGY, EDUCATION AND TRAINING, FELLOWSHIPS AND SCHOLARSHIPS

ALFRED O. OWINO
AND
JOSEPH O. OYUGI

Nova Science Publishers, Inc.
New York

LIBRARY OF CONGRESS CATALOGING-IN-PUBLICATION DATA

Owino, Alfred O.
Directory of conservation funding sources for developing countries : conservation biology, education & training, fellowships & scholarships / Alfred O. Owino, Joseph O. Oyugi.
p. cm.
Includes bibliographical references and index.
ISBN 978-1-60741-367-7 (hardcover : alk. paper)
1. Environmental agencies--Developing countries--Directories. 2. Conservation of natural resources--Developing countries--Finance--Directories. 3. Environmental protection--Developing countries--Finance--Directories. 4. Charitable uses, trusts, and foundations--Directories. I. Oyugi, Joseph O. II. Title.
GE20.O95 2009
333.7079--dc22 2009010240

Published by Nova Science Publishers, Inc. ✦ New York

CONTENTS

PREFACE

This directory provides an in-depth listing of potential donors for project and proposal developments. This is especially important considering that many students, organizations and individuals, especially in Africa, still lack adequate information on potential donor organizations to support their projects and career developments. Due to this problem, many good ideas with a high potential of contributing useful information for environmental conservation and development actions fail to get support from potential donors, either through lack of information or because of the type of approach strategies adopted. This directory will better help project and proposal developers to acquire information about donors that is necessary to target project ideas, determine a potential donor's interest in supporting or investing in a project, and eliminate donors whose missions and objectives do not match particular programmes. Once a match is made, this directory will also help to tailor project ideas to the specific requirements and interests of the donors.

"Nothing is impossible; there are ways that lead to everything, and if we had sufficient will we should always have sufficient means. It is often merely for an excuse that we say things are impossible" – Francois La Rochefoucauld.

RECOMMENDED CITATION

Owino, A.O. and Oyugi, J.O. (Ed.) (2009). Directory of Conservation Funding Sources for Developing Countries: Conservation Biology, Education and Training, Fellowships and Scholarship Awards. Nova Science Publishers, Inc, New York, USA.

DISCLAIMER

This publication contains the views of the editors in their individual capacities. The contents of this publication do not necessarily reflect the official views of the organizations mentioned or the grant institutions. The designations employed and the presentations do not imply the expression of any official position whatsoever on the part of organizations, the grant institutions or individuals concerning the legal status of any country, territory, city or area in its authority, or concerning the delimitation of its frontiers or boundaries. This publication does not legally bind the editors/compilers or contributors whatsoever to the users of the publication in terms of responses (negative or positive), comments or suggestion from the potential donors, organizations or grant institutions mentioned regarding any decisions whatsoever that an organization/institution may at their own discretion decide when approached for support, donations or inquires regarding their operations. This publication may be cited in whole or in part and in any form applicable for nonprofit purposes, but permission and of the copyright holder and acknowledgment is desired. It is further noted that some organizations mentioned herein may no longer exist or may have changed their priorities.

SUMMARY

Over the past five decades, man has changed ecosystems more rapidly than in any comparable period of time in human history, largely to meet growing demands for food, fresh water, timber and fuel (Mercer 2007). These changes are leading to serious deterioration of natural resources, with 60% of ecosystem services now degraded or used unsustainably (Millennium Ecosystem Assessment 2005). The developing countries and countries whose economies are in transition face numerous environmental and biodiversity conservation challenges. Overlaying the map of poverty on the environmental map reveals that throughout tropical and sub-tropical regions, the poor are concentrated in many areas that also have the greatest concentrations of biodiversity, and critical ecosystems that provide water and other resources (Mercer 2007). In many instances, funds/grants allocated by governments in developed and developing countries (see World Bank 2008) for environmental and conservation activities are often inadequate leading to high levels of dependence on funds/grants from other institutions or organizations.

For developing countries, most of support for environmental programmes and donor-based sourced from funding organizations based in developed countries. It is however, realized that 'donor fatigue' could be emerging as most funding organizations and environmental charities experience financial downturns leading such organizations to prioritize their programmes. For example, the New Philanthropy Capital (NPC) has found out that less than 2% of the annual grant-making of the 100 largest UK charitable trusts - £18.1m out of £1.1bn is allocated to environmental causes (Mercer 2007). With exception of a handful of foundations in the US and UK, the new philanthropy has been overwhelmingly focused on human and welfare issues (Charities Aid Foundation 2006). Grants/donor applications have also become very competitive, and there is always need for uniqueness in grant/donor proposals and clear justification of support for the proposed project. Given these patterns and scenarios, it is therefore appropriate for the donor/grant recipients to develop appropriate exit strategies for funded programmes to ensure that the projects/programmes continue to provide the desired benefits at the expiry of the donor/grant support period. This is important both to the grant/donor recipients as well as the institutions/organizations giving the donation or grants.

It is thus, important before embarking on grant or proposal writing are to ensure that the proposed activities can appeal to the potential donor or grant institution/organization. An in-depth research and making relevant inquiries into potential donors/grants is essential as it ensures that the project proposals are a good match to the funding organization, and conceptualized in such a manner that gives wider appeal and interest to the potential grant/donor. However, this process can be arduous with no guarantee that a funding proposal will receive funding once you have put in the time and effort in writing the grant proposal (Orlrich 1996). This problem is compounded further in developing countries where students, organizations and individuals still lack broad information on potential grants to support their projects proposals as there is still over-reliance on a handful of funding organizations, which does little to encourage confidence. As a result, environmental conservation and research ideas with high potentials of contributing useful information for environmental conservation, development actions and capacity building in the fast changing world fail to get much needed grant support. As identified by NPC, information, analysis and guidance are essential ingredients. Yet right across the environmental donor landscape, further progress is being constrained by lack of information (Mercer 2007).

It is with this regard that we have developed this comprehensive directory for conservation grant sources for developing countries to enhance information dissemination about the existing funding organizations. The Directory is aimed at contributing summarized information about the potential grant in the biological, ecology and conservation research, in addition to education, training and fellowship awards that are most relevant to developing countries. It provides summaries about the institutions/organizations and gives the relevant contacts including websites that can be accessed for additional information. This publication offers useful guides that are most relevant to project proposal developers. We provide some useful tips on grant application processes, project proposal budgets (adapted from different sources), and a selection a useful publications on fundraising that would be ideal for those keen in enhancing their skills in resources mobilization for conservation and environmental projects. In addition, we provide a comprehensive list of websites (assessed between 2006 and 2008) that give for detailed information about the relevant donors.

How to Use This Directory

This publication is divided into four broad sections: (i) Broader research, socioeconomic and development programmes; (ii) Education, training, fellowships and scholarships; (iii) Ecological, environmental and conservation research; and (iv) General sources. Although a considerable number of donors have cross-cutting programmes in the four sections, users of this publication would find valuable information in relevant areas of interest. For each funding organization, we provide summarized information and relevant contacts. Users of this publication are advised to use the contacts provided to seek further information from the respective websites and contacts. It is however, desirable to make an initial and very preliminary list of funding sources that seem to match your interests of your project. After making this selection, investigate further the interests and mission of the organization to be approached. This requires careful reading of internet sites provided. If possible, telephone or write to the organization directly and request for additional information as necessary. In particular, look for information about the potential donor/grant, subject interests and geographic focus. In many cases, the type of donor determines the approach to be adopted. The next stage should be to assess the type of programmes/projects that the donor supports as these often change. In addition, assess the general funding pattern of the donor especially the number of grants awarded annually including the funding ranges, organizations/projects the donor has funded in recent times, and relevant restrictions, specific application processes and formats. Only apply when satisfied that you meet the conditions outlined by the donor (Summarized from: The WorldFish Centre, 2008).

Grant Application Process: Some Useful Tips

Keep records of applications. It looks bad if a foundation/organization has been approached before, but you have no record – there is a loss of institutional memory when fundraising staff leave. Work on only a few as you must be able to keep track of the donor's interests, development of relationship-web with them and other donors on a particular project, names of key staff, etc, in case they phone. Alternatively have an efficient way of accessing your records, fast and following the donors' guidelines. Many donors now give operational support – they are realistic about this need and some are suspicious if you don't at least mention how you cover this. However, many will still give only program support. Tailor your application and budget accordingly. If possible, have a contact name. Do not send paper-works before your preliminary inquiry is responded to. Keep track of the process and if possible, call or e-mail to ask whether your latter/e-mail has been received. Remember not to harass your potential donor with unnecessary telephone calls or e-mails, and always respect the deadlines and eligibility requirements (*Summarized from different sources*).

Proposal Budgets: Some Basic Principles

A proposal budget can be defined as financial plan for a project, including all project income and project expenses, for a specified period of time. In many cases this should include the time-span. The first thing grant officers are likely to look at in a project proposal is the budget. If this seems realistic, the next thing they look at is the summary, followed by the covering letter. It is often necessary that someone other than the proposal writer prepares the budget. This also needs to be a team effort drawn from all players in proposed project to ensure that all likely costs and expenditure are covered. A donor may have own particular requirements for preparing the budget. Follow these requirements carefully to have your proposal considered. Many donors like applicants to use specific formats in preparing budgets. If you don't know the cost, get estimates from relevant sources and be prepared to say how these figures were obtained, and justify your proposal budget accordingly. In working out the budgets, it is important to take into consideration the following factors:-

1. *Accountability:* Donors have to account to both their own finance departments and to their stakeholders, including other government departments, parliament and taxpayers. They are required to show that the money from grants is being spent correctly and transparently. In awarding funding, donors must therefore ensure that the recipient has the capacity to manage their finances. You have the opportunity to convey this by including a clear, accurate budget within the proposal.
2. *Transparency:* Include all costs that will be necessary to carry out the research, ensuring that those types of costs will be covered by the donor.

3. *Justified costing:* Be sure to justify your costing in your proposal. A well thought out financial plan helps to create confidence in the proposal generally. Give as detailed a breakdown of costs as possible so that the funder can properly assess the case for support.
4. *Value for money:* Make sure that what you are asking for is allowed within the regulations and keep in mind that the donor will be looking for value for money.
5. *Be prudent:* Due to the time it takes between preparing, submitting, getting accepted and starting, many of the figures will be out of date, factor this into your original budget. Donors will often not give contingency money, but it is possible to build it into the budget. Think about inflation and how is it going to impact the figures.
6. *Make sure you ask for enough money:* Don't submit a budget which is not going to be enough to complete your project. If you save money donors will very rarely ask for it back. Donors are usually happy for money left over to be carried forward. Money left over can be useful for follow-up activities which donors are often not prepared to fund.
7. *Learn by doing:* You can learn budgeting by doing. It isn't necessary to have a finance team within the institution. If you can prove that you have thought about it, have learnt about it, and are a capable manager, this will give donors comfort that you are capable of handling the budgeting and finances of the project.
8. *Institutional capacity:* Some donors will look for institutions that offer a combination of solid institutional management capacity such as the ability to manage budgets and finance, with exciting, policy-focused research. However, the research performed by larger institutions can sometimes be run-of-the-mill research of limited interest to donors.
9. *Partnerships between large and small:* Donors are often attracted by the exciting research being performed by smaller institutions, though they can be put off by a smaller institutions lack of capacity at handling finances and budgets. Smaller institutions should look to make partnerships with larger institutions that can provide these donors with comfort that the financial management can be carried out satisfactorily.

REFERENCES AND SELECTED PUBLICATIONS ON FUNDRAISING

Barbato, J. and Danielle, S. F. (2000). *Writing for a good cause: the complete guide to crafting proposals and other persuasive pieces for nonprofits*. New York: Simon and Schuster.

Bauer, D.G. (2003). *The 'How To' Grants Manual: Successful grantseeking techniques for obtaining public and private grants*. Oxford, U.K: Greenwood Publishing Group.

Brewer, E.W. (2001). *Finding Funding: Grantwriting from start to finish, including project management and internet use*. Thousand Oaks, CA: Corwin Press.

Brown, L.G. and and Brown, M. (2001). *Demystifying grant seeking: what you really need to do to get grants*. 4th edition. San Francisco: Jossey-Bass Publishers.

Burk, P. (2000). Thanks! *A guide to donor-centered fundraising*. Burlington, Canada: Burk and Associates, Ltd.

Burke, J. and Carol, A.P. (2000). *I'll grant you that: A step-by-step guide to finding funds,designing winning projects, and writing powerful grant proposals*. Portsmouth, NH: Heinemann.

Burnett, K. (2002). *Relationship fundraising A donor-based approach to the business of raising money (2nd Edition)*. San Francisco, CA: Jossey-Bass, Inc.

Campbell, B. (2000). *Listening to your donors*. San Francisco, CA: Jossey-Bass, Inc.

Carlson, M. and Clarke, C. (2000). *Team-based fundraising step by step*. San Francisco, CA: Jossey-Bass, Inc.

Carlson, M. (2002). *Winning grants step by step: the complete workbook for planning, developing, writing, successful proposals (2nd Edition)*. San Francisco: Jossey-Bass Publishers.

Charities Aid Foundation (2006). *United Kingdom Giving 2005/2006*. UK: Charities Aid Foundation.

Ciconte, B.K. and Jeanne G.J. (2001). *Fund raising basics: A complete Guide*. Sudbury, MA: Jones and Bartlett Publishers, Inc.

Clarke, A.C. (2001). *Storytelling for grantseekers: The guide to creative Nonprofit fundraising*. California, USA: Chardon Press.

Coley, S. M. (2000). *Proposal writing*. Thousand Oaks, CA: Sage Publications.

Connors, T.D. (2001). *The nonprofit handbook (3rd Edition)*. New York: John Wiley and Sons, Inc.

Dove, K.E. (2001).*Conducting a Successful fundraising program: A comprehensive guide and resource*. San Francisco, CA: Jossey-Bass, Inc.

Dove, K.E., Vicky, L. M., Kathy, K.W., Mary, M.B. and Beggs, S.C. (2001). *Conducting a successful development services program.* San Francisco, CA: Jossey-Bass, Inc.

Duronio, M.A. and Eugene, T. (1996). *Fund raisers: Their careers, stories, concerns, and accomplishments.* San Francisco, CA: Jossey-Bass, Inc.

Flanagan, J.(1999). *Successful Fundraising: A complete handbook for volunteers and professionals.* New York, NY: McGraw-Hill Trade.

Geever, J.C. (2001). *The foundation center's guide to proposal writing.* NY: Foundation Center.

Gerding, S. and MacKellar, P.H. (2006). *Grants for libraries: a how-to-do-it manual.* New York: Neal-Schuman.

Golden, S.L. (1997). *Secrets of successful grantsmanship: A guerrilla guide to raising money.* SanFrancisco, CA: Jossey-Bass Publishers.

Grace, K.S. (1997). *Beyond fundraising: New strategies for nonprofit innovation and management.* New York: John Wiley and Sons, Inc.

Greenfield, J. (2001). *The nonprofit handbook: Fund Raising.* New York: John Wiley and Sons, Inc.

Greenfield, J.M. (1999). *Fundraising: Evaluating and managing the fund development process.* New York: John Wiley and Sons, Inc.

Hager, M.A. (Ed.) (2003). *New directions for philanthropic fundraising: Exploring measurement and evaluation efforts in fundraising".* San Francisco, CA: Jossey-Bass, Inc.

Hall, M.S. 1977. *Getting funded: A complete guide to proposal writing.* Portland: Portland State University.

Hall-Ellis, S.D. (2003). *Grants for school libraries*. Westport, Conn.: Libraries Unlimited.

Hayes, L., Hoffman, D. and Lamoreaux, D. (1999). *Winning strategies for developing grant proposals*. Washington, DC: Government Information Services.

Joyaux, S.P. (2001). *Strategic fund development: Building profitable relationships that last.* Sudbury, MA: Jones and Bartlett Publishers, Inc.

Karsh, E. and Arlen, S.F. (2003). *The only grant-writing book you'll ever need.* New York: Carroll and Graf.

Kelly, K.S. (1998). *Effective fundraising. Manuwah*, N.J.: Lawrence Erlbaum Associates.

Klein, K. and Stephanie, R. (Eds.) (2001). *Raise more money: The best of the grassroots fundraising journal.* San Francisco, CA: Jossey-Bass, Inc.

Klein, K. (2000). *Fundraising for social change (4th Edition).* California, USA: Chardon Press.

Klein, K. (2000). *Fundraising for the long haul.* Oakland, CA: Chardon Press.

Klein, K. (2001). *Fundraising for social change.* Oakland, CA: Chardon Press.

Klein, K. (2003). *Fundraising in times of crisis.* California, USA: Chardon Press.

Klein, K. (2006). *Fundraising for social change.* San Francisco, CA: Jossey-Bass Publishers.

Larissa G.B., Martin, J.B. and Nichols, J.E. (2001). *Demystifying grant seeking: What you really need to do to get grants.* San Francisco, CA: Jossey-Bass.

Levy, B.R. and Cherry, R.L. (Eds.). (1996).*The NSFRE fundraising dictionary*: NSFRE/Wiley fund development series. New York, NY: John Wiley and Sons, Inc.

Liane R.L. (2004). *Grant Application Writer's Handbook.* Boston: Jones and Bartlett Publishers.

Marshall, M.I., Aaron, J. and Fulton, J. (2006). *Writing a successful grant proposal.* Purdue University: Cooperative Extension Service.

Mcilnay, D.A. (1998). *How Foundations Work: What grantseekers need to know about the many faces of foundations.* San Francisco: Jossey-Bass Publishers.

McKinnon, H. (1999). *Hidden gold: How monthly giving Will build donor loyalty, boost your organization's income, and increase financial stability.* Chicago, IL: Bonus Books, Inc.

Mercer, B. (2007). *Green philanthropy: Funding charity solutions to environmental problems – a guide for donors and funders.* UK: New Philanthopy Capital Report.

Millenium Ecosystem Assessment (2005). *Millenium Ecosystem Assessment.* 5 volumes, plus synthesis reports for decision-makers.

Miller, P. W. (2008). G*rant writing: strategies for developing winning government proposals (3rd Edition).* Munster: Miller and Associates.

Miller, P.W. (2002). *Grant-writing: strategies for developing winning proposals.* Munster: P.W. Miller.

Miner, L.E , Miner, Miner, J.T. and and Griffith, J. (1998). *Proposal planning and writing (2nd Edition).* Milwaukee: Oryx Press.

Mixer, J.R. (1993). *Principles of professional fundraising: Useful foundations for successful practice.* San Francisco, CA: Jossey-Bass, Inc.

New, C.C. and Quick, A.J. (1998). *Grantseeker's toolkit : A Comprehensive Guide to Finding Funding.* New York: John Wiley and Sons.

New, C.C. and Quick, J.A. (2003). *How to write a grant proposal.* Hoboken, NJ: JohnWiley and Sons.

Nichols, J.E (1999). *Transforming fundraising: A practical guide to evaluating and strengthening fundraising to grow with change.* San Francisco, CA: Jossey-Bass, Inc.

Orlich, D.C. (1996). *Designing successful grant proposals.* England: Assn for Supervision and Curriculum.

Pricewaterhousecoopers, L.L.P, Michael B. K., Evelyn M. C. and Richard S. W. (2002). *Guide to charitable giving.* New York: John Wiley andSons.

Quick, J.A. (2001). *Grant seekers budget toolkit.* New York: John Wiley.

Rich, P. and Dana, H. (2002). *Membership development: An action plan for results.* Sudbury, MA: Jones and Bartlett Publishers, Inc.

Robinson, A. (2004) *Grassroots grants: An activist's guide to grantseeking (2nd Edition).* California: Chardon Press.

Robinson, E.M.M. and Klein, K. (Ed.) (2003). *The nonprofit membership Toolkit.* San Francisco, CA: Jossey-Bass, Inc.

Rosso, H.A (1996). *Rosso on fund raising: lessons from a master's lifetime experience.* San Francisco, CA: Jossey-Bass, Inc.

Rosso, H.A. and Associates and Eugene R.T. (Ed.). (2003). *Achieving excellence in fund raising (2nd Edition).* San Francisco, CA: Jossey-Bass, Inc.

Roth, S. and Ho, M. (2005). *The Accidental fundraiser: A step-by-step guide to raising money for your cause.* San Francisco: Jossey-Bass Publishers.

Scanlan, E.A. (1997). *Corporate and foundation fund raising: A complete guide from the inside.* Frederick, MD: Aspen Publishers.

Schaff, T. (1999).*The Fundraising planner: A working model for raising the dollars you need.* San Francisco, CA: Jossey-Bass, Inc.

Schumacher, E.C. (2003). *Building your endowment.* New York: John Wiley and Sons, Inc.

Seiler, T.L. (2001). *Developing your case for support.* San Francisco, CA: Jossey-Bass Publishers.

Seiler, T.L. (Ed.) (1998). *Understanding and improving the language of fundraising.* San Francisco, CA: Jossey-Bass, Inc.

Seiler, T.L. (Ed.) (2001). *Developing your case for support: Excellence in fund raising workbook series.* The fund raising school at the center on philanthropy at Indiana University. San Francisco, CA: Jossey-Bass, Inc.

Tempel, E. R. (Ed.) (2002). *Understanding donor dynamics: The organizational side of charitable giving.* San Francisco, CA: Jossey-Bass, Inc.

The WorldFish Center (2008). *Compendium of funding opportunities for Africa to support research, education, and development projects.* Penang, Malaysia: WorldFish Center.

Trenbeth, R.P (1993). *The membership mystique.* Rockville, MD: The Fund Raising Institute.

Wagner, L. (2001). *Careers in fundraising.* San Francisco, CA: Jossey-Bass, Inc.

Warner, I.R. (2001). *The art of fund raising.* Lincoln, NE: iUniverse, Incorporated.

Warwick, M. and Hitchcock, S. (2001). *Ten steps to fundraising success: choosing the right strategy for your organization.* San Francisco, CA: Jossey-Bass, Inc.

Warwick, M. (1999). *Five strategies for fundraising Success: A Mission-Based Guide to Achieving Your Goals.* San Francisco, CA: Jossey-Bass, Inc.

Weinstein, S. (2002). *The complete guide to fundraising management.* New York: John Wiley and Sons, Inc.

Weinstein, S. (2002). *The complete guide to fundraising management. (2nd Edition).* New York: John Wiley and Sons, Inc.

Weisman, C. (Ed.) (2000). *The secrets of successful fundraising.* St. Louis, MO: F.E. Robbins and Sons Press.

Ziarnik, N.R. (2003). *School and public libraries: developing the natural alliance.* Chicago: American Library Association.

World Bank (2008). *World Bank*, 7/1/07. USA: World Bank.

ACKNOWLEDGEMENTS

Compilation of this directory has been a long process. Many individuals and organizations have contributed to the information contained in this publication. We are very grateful to individuals and organizations – too many to mention here for there valuable information including individual funding organizations that were consulted but wished to remain anonymous. Additionally we are heavily indebted to a number of individuals who provided inputs and comments after taking time and care to read the different versions of this publication.

Section 1

BROADER RESEARCH, SOCIO-ECONOMIC AND DEVELOPMENT PROGRAMMES

A.W. MELLON FOUNDATION

This is a private foundation that makes grants on a selective basis to institutions working in the area of conservation and environment issues, among others. The Conservation and Environment Programme is devoted to basic research on how natural ecosystems work focusing on plant and terrestrial ecosystems. A small number of grants are devoted to research on how environmental regulations work and are used in actual practice.

Contacts: 140 East 62nd Street, NY 10021, New York – USA
Phone: +1-212-838-8400
wr@mellon.org or dnt@mellon.org
http://www.mellon.org/

ABERCROMBIE AND KENT GLOBAL FOUNDATION

Abercrombie and Kent is renowned as the world leader in luxury travel. The AandK Global Foundation promotes wildlife research and conservation as well as community education and historical monument preservation. Through active involvement in ongoing conservation projects, the Foundation helps to protect the delicate ecosystems in areas that are visited by A and K's guests. The goal of the AandKGF is to provide long-term support for existing natural ecosystems to ensure that human populations, flora and fauna continue to survive and coexist successfully.

Contacts: *1520 Kensington Road, IL 60523 Oak Brook – United States*
Phone: +1-630-472.1110; Fax: +1-630-471.1112;
info@akglobalfoundation.org
http://www.akglobalfoundation.org/home.html

Action for Greening Sahel

AGS operates in Burkina Faso and Chad in the Sahel region. Together with the people of the Sahel, it is working to stop desertification, the loss of any more green, and then to restore the natural ecosystem, allowing the people of the region to again be self-sufficient in their production of food. The organization provides funding and technical assistance for: the growing and distribution of seedlings; the provision of nursing advice and the installation of nursery facilities; stabilizing dunes by reforestation and stonework; installation of wells; and training in agricultural techniques.

Contacts: *Secretary, Nihon Sozo-Kyoiku-Kenkyujo Bldg. 3F, 2-2-7, Minami-Suna, Koto-ku, 136-0076 Tokyo, – Japan; Phone:+81-3-5632-3029, Fax: +81-3-5632-3070*
sahel@jca.apc.org
http://www.jca.apc.org

Action for Mangrove Reforestation

ACTMANG was founded in 1992 with the goal of promoting the preservation and reforestation of the mangrove ecosystems. Model projects for mangrove reforestation have been underway in Vietnam, Ecuador and Myanmar.

Contacts: *MM Mansion 1104, 3-29-15-1104, Honcho Nakano-ku,*
164-0012, Tokyo – Japan
Phone: +81-3-3373-9772; Fax: +81-3-3373-9772
actmang@big.or.jp

Action for World Solidarity

Since 1957, The Action for World Solidarity Organization (ASW) has supported small, self- initiated groups in India, Africa and Brazil, which try to strengthen and improve the situation of women, strive for environmental protection, or attempt to achieve social and cultural human rights. Action for World Solidarity promotes projects which have already begun working, proven their effectiveness and need further financial assistance in order to continue functioning. Instead of sending experts or development helpers abroad, we leave the responsibility of running the projects entirely to those people who have initiated them. ASW supports small projects over a number of years in order to allow them time to become established and capable of securing their own financial sources.

Contacts: *Aktionsgemeinschaft Solidarische Welt, ASW (Action for World Solidarity); Hedemannstraße 14; 10969 Berlin Germany*
Phone: ++49 / 30 / 25 94 08 01, Fax: ++49 / 30 / 25 94 08
Email: mail@aswnet.de
http://www.en.aswnet.de

AEON FOUNDATION

AEON is involved anti-desertification measures and recycling of resources, and is also addressing the problems of waste disposal through own environmental conservation activities and supporting the work of other groups with similar objectives. Activities which are supported are tree planting, greening, provision of water and ecology preservation, but is also involved in global environmental conservation activities. Applicants should be endorsed by individual(s) or group(s) in Japan (learned persons familiar with environmental issues or experts in appropriately related fields are considered apt as presenters.

Contacts: *5-1, 1-chome, Nakase, Mihama-ku, Chiba-shi, 261-8515 Chiba – Japan*
Phone:+81-43-212.6022; Fax: +81-43-212.6815
ef@aeon.info or aeon-ef@mailgw.jusco.co.jp
http://www.aeon.info

AFRICAN DEVELOPMENT FOUNDATION

The United States Congress established the African Development Foundation (ADF) in 1980 as an independent public corporation with a mandate to promote the participation of Africans in the economic and social development of their countries. The principal aim of ADF grants is to enable grassroots groups to generate increased incomes through productive enterprises that expand the overall economic production capacity and increase the economic security of their families and communities. USADF provides small grants to private and other nongovernmental entities in Africa to: finance sustainable poverty alleviating initiatives that are conceived, designed, and implemented by Africans and aimed at enlarging opportunities, for community development, stimulate and expand the participation of Africa's poor in the development of their countries, and build sustainable African institutions that foster grassroots development.

All USADF-funded projects are "made in Africa" - proposed, designed, and implemented by the beneficiaries themselves, with any needed technical assistance and training provided entirely by local experts. Legally registered cooperatives, economic interest groups, producer associations, legally registered limited liability companies, and community-based development organizations based in African countries where USADF works are eligible, and projects should be led by citizens of the country where the project would be based. All questions regarding grant eligibility or grant application procedures should be directed to the ADF country representative who administers the country program in the country where your project would be based.

Contacts: *African Development Fund; 1400 I Street NW, 10th, Floor;*
Washington D.C. 20005-2248, USA
Phone: (202)673-3916, Fax: (202)673-3810
info@adf.gov
http://www.adf.gov/funding.html

African, Caribbean and Pacific Countries (ACP)/ (DG VIII)

Citizenship of ACT country. Applicants must be nominated by their own government - Check website for details and deadlines.

Contacts: *Directorate General VIII (DG VIII),*
Commission of the European Communities, Rue de la Loi 200, 1049 Brussels, Belgium
Phone. +32 2 743 06 00 or 32 2 299 1111 (Tel. only)
info@acpsec.org
http://www.acpsec.org

Africare

Africare initially concentrated on helping to alleviate the effects of severe drought in West Africa. By the mid-1970s, Africare shifted its emphasis to development programs in the areas of food, water, the environment and health. The organization is active in 24 countries of Sub-Saharan Africa and in Egypt. Africare's programs address needs in the principal areas of food security and agriculture as well as natural resource management. Africare also supports water resource development, environmental management, basic education and vocational training, micro-enterprise development, civil-society development and governance initiatives. Africare helps farmers in dry, drought-prone to combat soil's erosion and to cultivate crops and nurture their livestock despite scarce water supplies.

Contacts: *Francophone West and Central Africa); Kevin G. Lowther and Sharie A. Blanton (Southern Africa Region). Address: Africare House, 440 R Street, N.W., DC 20001 Washington - United States.*
Phone: +1-202-462.3614; Fax: +1-202-387.1034
development@africare.org
http://www.africare.org

Aga Khan Foundation

The Aga Khan Foundation (AKF) is a non-denominational, international development agency established to develop and promote creative solutions to problems that impede social development, primarily in Asia and East Africa. The Foundation focuses on a small number of specific development problems by forming intellectual and financial partnerships with organizations sharing its objectives. The AKF funds projects that within a cluster of inter-related programs seeking to resolve generic problems in health, education, rural development, or the enhancement of nongovernmental organization capacity. Most Foundation grants are made to grassroots organizations testing innovative approaches.

Contacts: *Aga Khan Foundation; 1-3 Avenue de la Paix; P.O. Box 2369; 1211*
Geneva 2, Switzerland
Phone: +41 22 909 7200, Fax: +41 22 909 7291
akfgva@atge.automail.com
http://www.akdn.org/agency/akf.html

AGFUND International Prize

The Arab Gulf Programme for United Nations Development Organizations (AGFUND) is a regional developmental institution, established in 1980 upon the initiative of HRH Prince Talal Bin Abdul Aziz, and with the support of the Leaders of the Arab Gulf States that constitute its membership and contribute to its budget. AGFUND is concerned with the support of sustainable human development efforts, targeting the neediest groups in the developing countries, particularly women and children, in cooperation with the organizations and institutions active in this field. AGFUND International Prize is an annual prize awarded by the Arab Gulf Programme for United Nations Development Organizations (AGFUND), having as its president HRH Prince Talal Bin Abdul Aziz Al Saud, to honor the pioneering development projects in the developing country. It aims at encouraging continuity in implementing pioneering development projects and stimulating the efforts of those who are capable of contributing to developmental and humanitarian action in the developing countries. This reflects the keenness and continued endeavors of AGFUND to support the efforts targeting the alleviation of human suffering and pain.

The United Nations organizations, international and regional organizations, universities, research centers, NGOs, Governmental bodies (if not participating in the project implementation) and country federations of NGOs. What is eligible: Pioneering projects implemented by UN, international or regional organizations; Pioneering projects implemented by nongovernmental organizations; and Pioneering projects founded, sponsored and/ or implemented by individuals. AGFUND Prize is classified into three categories in accordance with the nature of the implementing bodies of the nominated projects.

Contacts: *P.O. Box 18371; Riyadh 11415; Saudi Arabia*
Phone: +966 1 4418888, Fax 1: +966 1 4412962, Fax 2: +966 1 4412963
Email: info@agfund.org or prize@agfund.org
http://www.agfund.org

Aktionsgemeinschaft Solidarische Welt

Aktionsgemeinschaft Solidarische Welt means Action Community for a Solidarity World; solidarity in cooperation is the central issue for the ASW. ASW finances projects in the areas of environmental protection, women's empowerment, food sovereignty and human rights. ASW especially supports groups in the South, which want to ameliorate their living conditions based on alternative concepts and (small-scale) local initiatives. Its Environment Fund focuses on ecology and sustainability.

Contacts: *Hedemannstr. 14, 10969 Berlin – Germany, Phone: +49-30-251.02.65;*
Fax: +49-30-251.18.87
E-mail: mail@aswnet.de
http://www.aswnet.de

ALEXANDER ABRAHAM FOUNDATION

The Alexander Abraham Foundation is a not-for-profit organization with an environmental and conservation focus. AAF is dedicated to conserving the environment without boundaries and to defending indigenous wildlife, with an emphasis on protection of endangered species. AAF pursues its mission at the local level by establishing relationships with partner organizations that work in the field and on the ground in Central and Western Africa, and in Asia. AAF also provides consulting services that enable grantees to become self-sustaining and improve their ability to promote their own missions. The Foundation's primary areas of interest are: environment and wildlife protection; animal welfare; ecotourism; Abraham Conservation Awards: The Abraham Award focuses on honoring grassroots activists in whatever area of environmentalism they make a significant contribution. In addition to public recognition, winners receive a cash award and certificates. The recognition through these awards is also meant to inspire, sustain and motivate future conservation efforts in the region. The Foundation has supported dozens of projects around the world, with partners ranging from the World Wildlife Fund and the Wildlife Conservation Society, to non-profits staffed by a handful of scientists. Check the website for updates and deadlines.

Contacts: *The Alexander Abraham Foundation; 232 East 62nd St.;*
New York, NY 10021 USA
Phone: (646)217.4207, Fax: (212)521.0799
info@aabrahamfoundation.org
http://www.aabrahamfoundation.org/index.htm

ALLIANCE FOR INTERNATIONAL REFORESTATION

The objective of this small organization is to assist local community groups in Central America to conserve their environment through reforestation, sustainable farming, and education. AIR plants trees, establishes tree nurseries, digs wells, builds fuel-efficient brick ovens, and helps to educate everyone about the environmental challenges facing Central America. The establishment of permanent tree nurseries is the keystone of AIR's work. The aim is to train community groups to initiate a process of reforestation and conservation of their land. In order to accomplish this, AIR assists community groups to organize reforestation committees and provides them with educational and training programs designed to increase their ability to establish, manage and maintain tree nurseries. These groups are directly responsible for implementing project activities, such as planting seeds, caring for the saplings, maintaining the nurseries and transplanting the trees. AIR's staff provides technical assistance as needed. All of the equipment needed for the construction and maintenance of the nurseries is provided by AIR, while the communities contribute the labor in a voluntary fashion.

Contacts: *Stetson University, 421 North Woodland Boulevard, Unit 8301,*
FL 32723 DeLand - United States
Phone: +1-386-822.7575;
http://www.stetson.edu/org/forest

American Jewish World Service

AJWS is a non-profit, international development organization working on a nonsectarian basis with grassroots NGOs throughout the developing worlds that are engaged in community development or emergency relief. The inspiration for AJWS' work is drawn from the mandate to work for social justice that is articulated in the Jewish tradition. AJWS provides financial support in the form of small grants, awarded for a year at a time; and may be renewed annually. AJWS funds projects in the areas of sustainable agriculture, health, economic development and education. All projects should have a strong component of strengthening civil society or women's empowerment. Grassroots NGOs only are eligible.

Contacts: *Department of International Programs; American Jewish World Service; 45 West 36th St.; New York, NY 10018 USA, Fax: (212) 736-3463,*

grants@ajws.org

http://www.ajws.org

Ancient Forests International

AFI is dedicated to the protection of primary forests around the world by developing long-term, viable projects to protect some of the world's last wild places. Its main activity is the Acquisition to acquire and protect strategic and invaluable forestlands. AFI has helped coordinate the purchase of nearly a million acres of ecologically critical forested land, primarily along the Pacific coast of North and South America. AFI is also helping to support the neighboring communities in stewarding those areas for generations to come. AFI supports efforts to develop locally appropriate economic strategies to support the people who live in or near core protected lands, in the interest of protecting those places over the long-term.

Contacts: *PO Box 1850, CA 95560, Redway – USA*

Phone: +1-7070-923.3015; afi@ancientforests.org

Arab Science and Technology Foundation: Abdul Latif Jameel Grants

The ALJ Grant program aims at enhancing civilian research and development activities inside the Arab world towards outputs that contribute to the sustained social and economic development. Emphases are placed on research proposals in medical, engineering, applied sciences, and other sciences that are meeting the following requirements: provide appropriate technologies for developing the Arab world communities, develop affordable products through designs, materials and mechanisms whilst building a suitable business plans and strategies that benefit the Arab world communities. Address target markets including, but not limited to, energy, medical care, water purification, environmental hazardous, telecommunication, food safety, plant improvement and breeding, household equipment, and others. Furthermore, one third of the ALJ Grant is to be utilized on projects that address the problems of the poor.

Contacts: *http://www.astf.net/site/index.asp*

ASHDEN AWARDS FOR SUSTAINABLE ENERGY: INTERNATIONAL AWARDS

The International Awards are designed for schemes in the developing world. Award winners use local renewable energy to reduce poverty, improve people's health, well-being and economic prospects, and at the same time tackle climate change and other environmental threats, notably deforestation. Prizes are awarded for schemes which address at least one of the following areas: (i) food security, (ii) health and welfare, (iii) light, (iv) education, and (v) enterprise. One of the five awards take the form of a special African Award, reserved specifically for an outstanding scheme from that continent. NGOs, businesses or, exceptionally, individuals are eligible.

Contacts: *The Ashden Awards for Sustainable Energy; Allington House; 150 Victoria Street; London SW1E 5AE UK*
Phone: +44 (0)20 7410 0330,
info@ashdenawards.org
http://www.ashdenawards.org

AUSTRALIAN AGENCY FOR INTERNATIONAL DEVELOPMENT

AusAID is responsible for the management of the Australian government's overseas aid program. NGOs are able to seek funding within individual country programs in developing countries. Substantial aid is provided to African countries, but most of the aid assists the Asia-Pacific region. All aid projects are designed to promote development in ways that are environmentally sustainable.

Contacts: *GPO Box 887, ACT 2601 Canberra – Australia.*
Phone: +61-2-6206.4000; Fax: +61-2-6206.4880;
InfoAusAID@ausaid.gov.au
http://www.ausaid.gov.au

AUSTRALIAN CENTER FOR INTERNATIONAL AGRICULTURAL RESEARCH

ACIAR encourages Australia's agricultural scientists to use their skills for the benefit of developing countries and Australia, and funds research projects that are developed within a framework reflecting the priorities of Australia's aid program and national research strengths, together with the agricultural research and development priorities of partner countries. The mandate is to direct activities to developing countries in five regions: Papua New Guinea and the Pacific Islands, Southeast Asia, North Asia, South Asia and Southern Africa. Research is also allocated across regions through funding to the international agricultural research centres. For Africa, new projects considered are on beef production, crop–livestock systems (involving small and large ruminants) and the efficient use of nutrients and water in cropping systems. Small grant proposals should be submitted to ACIAR prior to the first calendar year of proposed funding.

Contacts: *The Director, ACIAR; GPO Box 1571; Canberra ACT 2601 Australia, Phone: (02) 6217 0500 (Int. Code 612), Fax: (02) 6217 0501 (Int. Code 612), aciar@aciar.gov.au (for information and project requests), http://www.aciar.gov.au*

BARING FOUNDATION: JOINT INTERNATIONAL DEVELOPMENT GRANTS

The Baring Foundation has specific grants programmes concerned with strengthening the voluntary sector, the arts and international development. A recently funded project in international development was: Pastoral and Environmental Network in the Horn of Africa (PENHA). The programme guidelines for the year are announced on their website 2-3 months prior to the next deadline date.

Contacts: *The Baring Foundation; 60 London Wall; London EC2M 5TQ UK, Phone: 020 7767 1348, Fax: 020 7767 7121, baring.foundation@uk.ing.com http://www.baringfoundation.org.uk*

BELGIAN SURVIVAL FUND

The main objective of the Fund is to improve the food security of the most vulnerable population groups in the poorest countries. The focus of the BSF is on the countries of sub-Saharan Africa, and especially on the partner countries of direct bilateral cooperation. The BSF finances programmes in countries faced with a chronic food shortage, in order to guarantee the chances of survival of persons threatened by hunger, under-nourishment, poverty and exclusion. The programmes favour an integrated approach with the aim of improving the food and nutritional security of families and local communities in rural and semi-urban environments. The Fund also finances programmes whose aim it is to reinforce the institutional capacities of the national and local authorities and the local communities being assisted. The operations of the Survival Fund are carried out within the framework of the partnership programmes with 4 multilateral organizations (FAO, IFAD, UNICEF, and UNCDF) and 14 Belgian NGOs.

Contacts: *Programme officer, Rue des Petits Carmes 15, 1000 Brussels – Belgium, Phone: +32-2-501.81.11, http://www.dgdc.be/en/actors/belgian_survival_fund/index.html*

BILL AND MELINDA GATES FOUNDATION

The Bill and Melinda Gates Foundation awards the majority of its grants to U.S.501 (c)(3) organizations and other tax exempt organizations identified by foundation staff according to the objectives of four program areas: Global Health, Education, Global

Libraries, and Pacific Northwest. The Foundation favors preventive approaches and collaborative endeavors with government, philanthropic, private sector, and not-for-profit partners. Priority is given to projects that leverage additional support and serve as catalysts for long-term, systemic change. Funding is directed according to several themes: global development program: agricultural development, financial services for the poor, global libraries, special initiatives. Global health programme: critical enabling technologies, community health solutions, emergency relief. United States Program: charitable sector support.

Contacts: *Bill and Melinda Gates Foundation; PO Box 23350; Seattle, WA 98102 USA, Phone: (206) 709-3100 (Reception); (206) 709-3140 (Grant Inquiries), info@gatesfoundation.org;*
http://www.gatesfoundation.org

BirdLife International Conservation Leadership Programme

The Conservation Leadership Programme aims to contribute to long-term environmental conservation and sustainable development in priority areas by encouraging and engaging potential leaders in biodiversity conservation, and providing opportunities for them to gain practical skills and experience. The programme offers continuation funding in the form of Conservation Follow-Up and Conservation Leadership Awards. Conservation Follow-Up Awards are aimed at exceptional teams with individualswho want to address a conservation issue raised by recommendations in a project previously supported by the programme. The project should allow team members to go beyond basic surveying, analysis, and awareness-raising, and develop their networking, decision-making, policy, communications, and conservation leadership skills and experience.

Contacts: *Conservation Leadership Programme c/o BirdLife International*
Wellbrook Court, Girton Road,
Cambridge CB3 0NA United Kingdom,
Phone +44 (0) 1223-279822; Fax +44 (0) 1223-277200
clp@birdlife.org

Blue Link

The Blue Link initiates, manages and participates in companies that show solid commercial returns and considerable development impact in emerging markets in developing countries. The Blue Link operates from offices in the Netherlands, Kenya and Pakistan. In the Pakistani Punjab, The Blue Link is developing a position in the dairy farming sector through the company ProFarm Pakistan. ProFarm sells high quality bovine semen to dairy farmers through the establishment of a sales and service infrastructure in the Punjab. The Blue Link is currently involved in setting up related businesses in this sector. In Kenya, The Blue Link is running the TBL Mirror Fund that invests in companies in the Kenyan Small and Medium Enterprises sector. The Blue Link is currently exploring to set up additional investment services for Sub Sahara countries. All enterprises of The Blue Link start with a clear market demand in the countries of operation. The Blue Link works with local partners to set up local entities that manage and develop new business. The Blue Link has a wide network of investors and experts available who are matched with the new businesses. Using its business development and management expertise. The Blue Link thus establishes and manages new, innovative and well networked companies.

Contacts: *The Blue Link BV; Arlandaweg 92; 1043 EX Amsterdam The Netherlands;*
Phone: +31 (0)20 4003129 or +31 (0)20 6845146;
Fax: +31 (0)20 4862256; info@thebluelink.org;

ProFarm Pakistan LTD; 69/7 Green Villas – FCC; Syed Maratib Ali Road;
Gulberg IV - Lahore 54660 Pakistan, Phone: +92 (0)42 5777846-7,
Fax: +92 (0)42 5777811; info@profarm.com.pk; http://www.profarm.com.pk;

TBL Mirror Fund BV; Gatundu-Kandara Road Junction; Kileleshwa, Nairobi;
P.O. BOX 79469 - 00200 Nairobi Kenya; Phone: +254(0)20 2019374;
Fax: +254(0)20 4343907; info@tblmirrorfund.com
http://www.tblmirrorfund.com

British American Tobacco Biodiversity Partnership

The British American Tobacco Biodiversity Partnership, launched in 2000, involves British American Tobacco and four conservation NGOs: Earthwatch Europe; Fauna and Flora International; the Royal Botanic Gardens Kew and the Tropical Biology Association. It aims to: work through external projects to contribute to the conservation of global biodiversity by building a portfolio of activities that assist countries where the partners operate in meeting their obligations under the Convention on Biological Diversity; and work within the British American Tobacco Group and amongst the other partners to enhance the partners' own ability to conserve and manage biodiversity.

Contacts: *http://www.batbiodiversity.org*

Bruno Manser Fonds

This fund is an association for the indigenous peoples of Malaysia, particularly the Penan people of Sarawak. This small fund programme focuses on protecting the remaining virgin forest of the Penan home land, and prioritize activities which are designed to halt logging by the timber industry and to raise awareness of the rights of the Penan people.

Contacts: *Heuberg 25, 4051 Basel – Switzerland*
Phone: +41-61-261.94.74; Fax: +41-61-261.94.73
info@bmf.ch
http://www.bmf.ch

C. S. Fund/Warsh-Mott Legacy

The foundation makes funding decisions twice a year, usually in April/May and November/December. If a full proposal is requested, it must be received by January for consideration in the spring, or in August for consideration in the fall. Applicant organizations must be classified as a 501(c) (3) by the US Internal Revenue Service. Foreign applicants should note that the foundations make a very limited number of grants abroad. The CS Fund and Warsh/Mott Legacy are private foundations which are linked by common issue interests and boards of directors. CSF and WML are dedicated to preserving biodiversity, defending democracy, preventing the commoditization of life, and protecting human and environmental health. Occasionally the foundation may initiate support for projects that fall outside their established priorities. These include but are not limited to efforts to advance the precautionary principle and the commons.

Contacts: *CS Fund; 469 Bohemian Highway; Freestone, CA 95472 USA;*
Phone: 707-874-2942, Fax: 707 874 1734
inquiries@csfund.org
http://www.csfund.org

Canadian International Development Agency

CIDA is the Canadian Government's development agency. In addition to supporting basic human needs, gender equality, and infrastructure services, the Agency also gives attention to issues of human rights, democracy, good governance and strengthening both civil society and the security of the individual. In the area of the environment, the Agency helps developing countries to protect their environment and to contribute to addressing global and regional environmental issues.

Contacts: *200 Promenade du Portage, Quebec, K1A 0G4 Hull – Canada*
Phone: +1-819-9975456; Fax: +1-819-9535469;
info@acdi-cida.gc.ca
http://www.acdi-cida.gc.ca

Canadian Lutheran World Relief

CLWR is the service delivery arm for development programming and overseas relief for the Evangelical Lutheran Church in Canada and Lutheran Church-Canada. The principle mission centres on development initiatives in partnership with the marginalized of the world community. CLWR facilitates and supports development programmes in Africa, Asia, Latin America and the Middle East. One of its areas of work is long-term development assistance. Special attention is paid to the integration of development projects with a sensitive (critical) concern for the environment with special attention for on the use of appropriate technology. Project priorities are: soil and water conservation (i.e. by curbing the rate of deforestation through afforestation campaigns); rehabilitation of irrigation canals and drinking water supplies; and agro-ecological development with the objective to increase sustainable agricultural production through ecologically-sound cultivation practices.

Contacts: *1080 Kingsbury Avenue, R2P 1W5 Winnipeg – Canada*
Phone: +1-204-694.5602; Fax: +1-204-694.5460;
clwr@clwr.mb.ca
http://www.clwr.org

Cargill Corporate Giving Programme

The Cargill Citizenship Fund provides strategic grants to organizations serving communities where Cargill has a presence. The company provides direct grants for regional, national and global partnerships and provides matching funds for selected local projects supported by their businesses. Cargill provides corporate support to organizations, programs, and projects that fall within the Committee's three focus areas: nutrition and health; education; environment. There are no deadlines, and applications are reviewed on an ongoing basis. Regional, national or global nonprofit or non-governmental organizations that provide programmes and services to multiple Cargill communities may apply directly to the Cargill Citizenship Fund. Local, community organizations must request funding from their Cargill Facility.

Contacts: *Cargill Citizenship Fund; Director, Corporate Contributions; P.O. Box 5650; Minneapolis, MN 55440 USA*
Phone: (952) 742-2931
http://www.cargill.com

Carnegie Corporation of New York Grants

The four program rubrics are: education, international peace and security; international development (applies to Africa) and strengthening U.S. democracy. Grant seekers keen to approach the foundation with a preliminary request for funding are encouraged to submit a letter of inquiry. If the project described in the letter fits the foundation's guidelines, the sender will be contacted and asked to submit a proposal in the Corporation's format. A request to submit a proposal is not an indication of the Corporation's intention or commitment to award a grant. The Corporation accepts requests for funding at all times of the year and welcomes inquiries from potential grantees whose work fits strategic guidelines. Carnegie Corporation does not make grants for construction or maintenance of facilities or endowments, and does not generally make grants to individuals except through the Carnegie Scholars Programme.

Contacts: *Carnegie Corporation of New York; 437 Madison Avenue, New York, NY 10022 USA*
Phone: (212) 371-3200; Fax: (212) 754-4073
http://www.carnegie.org/sub/program/index.html

Centrum für Internationale Migration und Entwicklung

CIM places managers and technical experts in Asia, Africa, Latin America, and Eastern and South Eastern Europe, and support them with services and with subsidies to top up their local salaries. The mission is to support partners' contributions to their countries' development and the attainment of the goals set jointly by each partner country and the German Federal Ministry for Economic Cooperation and Development (BMZ). This is accomplished by supplying the demand of these countries for the kind of highly-qualified professionals they could not attract under customary national employment conditions. CIM places Integrated Experts and Returning Experts Through the Integrated Experts Programme, CIM links up partner country organizations needing qualified employees with highly qualified experts from Germany and other European Union countries – quickly and unbureaucratically. By doing so, CIM paves the way for know-how transfer to developing, transition and emergent countries. CIM helps Returning Experts (individuals from developing, emergent, and transition countries live and work in Germany or have completed their education or training thereto) be more effective. In this way CIM helps make international migration processes sustainable in terms of employment and development policy.

Contacts: *Centrum für Internationale Migration und Entwicklung (CIM)*
Mendelssohnstrasse 75-77; 60325 Frankfurt am Main Germany
Phone: + 49 (0)69 719121 0; Fax: + 49 (0)69 719121 19
cim@gtz.de
http://www.cimonline.de/en/index.asp

CGIAR King Baudouin Award

In 1980, the Consultative Group on International Agricultural Research (CGIAR) received the King Baudouin International Development Prize from the King Baudouin Foundation "for its contribution to the qualitative and quantitative improvement of food production in the world." After consultations with the Kingdom of Belgium, and using funds received from the King Baudouin International Development Prize, the CGIAR established its own biennial King Baudouin Award to acknowledge and stimulate agricultural research and other activities relevant to the system and to recognize an achievement stemming from the work of a Center and its partners. The award is given only once every 2 years after close scrutiny by the Technical Advisory Committee of the CGIAR. Nomination is invited from each Center, in accordance with the following guidelines: (i) direct or indirect, actual or potential impact on resource poor farmer and low-income people; (ii) direct or indirect, actual or potential impact on sustainable production systems; (iii) innovation in science; (iv) partnership/collaboration.

Contacts: *http://www.cgiar.org/awards/awards_baudouin.html*

CHARLES STEWART MOTT FOUNDATION

The mission of the Fund's Environment program is to support the efforts of an engaged citizenry working to create accountable and responsive institutions, sound public policies, and appropriate models of development that protect the diversity and integrity of selected ecosystems in North America and around the world. The funding priorities that follow describe how the Foundation hopes to contribute to meeting these challenges. The Environment program is organized into three program areas: (i) conservation of freshwater ecosystems (limited to North America); (ii) international finance for sustainability goal (to shape international investment and trade to support sustainable; development and reduce environmental degradation), and (iii) special initiatives (through this program area, unusually compelling environmental projects that fall outside of our two core program areas are supported). The Mott Foundation makes grants in the U.S. and, on a limited geographic basis, internationally. Grants outside the U.S. are made only in the civil society and environment programs. Request for proposals from organizations which the Foundation has been in contact are sent, but unsolicited requests for projects that fall within our program priorities and guidelines are also accepted. Applications are accepted year-round from Non-governmental organizations (NGOs).

Contacts: *Office of Proposal Entry; C.S. Mott Foundation; Mott Foundation Building*
503 S. Saginaw St., Suite 1200; Flint, MI 48502-1851 USA
http://www.mott.org

CHURCH OF SWEDEN AID

Church of Sweden Aid is Church of Sweden's organization for development cooperation and emergency aid working to enhance human dignity in people's living conditions as well as fighting injustice, poverty and oppression. An important working field for the organization is sustainable ecological development, and ecological considerations are an important aspect of development projects.

Contacts: *751 70 Uppsala – Sweden,*
Phone: +46-18-169600; Fax: +46-18-169707
lutherhjalpen@svenskakyrkan.se
http://www.svenskakyrkan.se

CHRISTENSEN FUND

The Fund focuses grant making on the interface between natural environments and human cultures, primarily working with peoples and communities variously known as "indigenous", "tribal" and "minority", and the landscapes with which their cultures, histories and artistic expression are so connected. An objective is to back locally grounded efforts to elucidate the relationships between culture and environment and to strengthen the world's capacity to maintain cultural and biological creativity and diversity going forward. In addition to the focused work in the four selected regions, TCF will also fund some "Global Bicultural

Initiative") (formally known as "cross-cutting") efforts involving those regions and the wider world. Applicants seeking grants can submit an inquiry. Grants are directed to organizations based within the following regions The Greater South West (Southwest USA and Northwest Mexico), Central Asia and Turkey, The African Rift Valley (especially Southwest Ethiopia and also Northern Kenya), Northern Australia and Melanesia.

Contacts: *394 University Ave.; Palo Alto, CA 94301 USA*
Phone: (650)323 8730
info@christensenfund.org
http://www.christensenfund.org

COCA-COLA AFRICA FOUNDATION

The Coca-Cola Foundation, and the company's other 18 local and regional foundations provide the means to help improve the well-being of society. The Foundation works with people in communities and governmental organizations and nongovernmental organizations to create and support projects most relevant to communities. The programmes, which also include customized local initiatives, converge in the areas of the environment, sports and physical activity, education and disaster relief. In 2001, the Company created The Coca-Cola Africa Foundation as the engine to coordinate their corporate social investment in Africa. The foundation addresses individual and collective needs across issues of health, education and the environment, as well as supporting subject-matter expert development.

Contacts: *The Coca-Cola Africa Foundation; P.O. Box 2040; Manzini, Swaziland*
tccaf@eur.ko.com
http://www.thecoca-colacompany.com

COMPTON FOUNDATION

The Foundation was founded to address community, national and international concerns in the fields of peace, world order, population, and the environment. The Foundation's priorities in the environmental field include: land, river, and watershed protection and management for purposes of long-term habitat and ecosystem preservation and restoration; changing the relationship between people and the natural environment in order to promote sustainable and just balance between meeting present human needs and conserving natural systems for future generations. Priority is given to projects which more than local application, are replicable and are likely to be taken over and managed by long-term funding sources.

Contacts: *545 Middlefield Rd. Suite 178, CA 94025 Menlo Park - United States*
Phone: +1-650-328-0101; Fax: +1-650-328-0171
info@comptonfoundation.org
http://www.comptonfoundation.org

CONCERN AMERICA

Concern America recruits and sends professionals who serve as non-salaried volunteers for two or more years overseas. Medical doctors, public health specialists, educators, agriculturalists, engineers, and other experts are sent to share their skills and knowledge with community leaders. After eight to ten years in an area, the Concern America team leaves in place trained local people who are "capacitated" to continue the work and to bring it to neighboring communities. Concern America is amongst others active in development projects in the fields of agro-forestry, organic farming and fuel-efficient stoves.

Contacts: *PO Box 1790, CA 92702 Santa Ana – United States*
Phone: +1-714-953.8575; Fax: +1-714-953.1242
concamerinc@earthlink.net
http://www.concernamerica.org

CONRAD N. HILTON HUMANITARIAN PRIZE

Prize is awarded annually to an organization making extraordinary contributions toward alleviating human suffering anywhere in the world. Nominees must be organizations, not individuals, non-governmental, publicly supported charitable organizations. Nominees must be legally established in order to be considered, and must have an operating budget greater than U.S. $500,000

Contacts: *prize@hiltonfoundation.org*
http://www.hiltonfoundation.org

CORDAID

The Dutch Catholic development organization has joined the forces with Bilance, Memisa, Mensen in Nood (Caritas Netherlands) and Vastenaktie (Dutch Bishops' Lenten Campaign). In this way, it is involved in all aspects of development cooperation including emergency aid, structural poverty alleviation and health care. Cordaid supports projects and programmes from NGO's that involve the active participation of the local population and pursue tangible results for the individual beneficiaries, their self-organization and its influence on other local actors. In this development approach, special attention is given to environment, gender and geographical and sectoral coherence. The organization receives its money from the Dutch government, the European Union and financial donations from individual contributors who support the work of Memisa, Mensen in Nood, and Vastenaktie.

Contacts: *Postbus 16440, 2500 BK Den Haag – Netherlands*
Phone: +31-70-3136300; Fax: +31-70-3136301
cordaid@cordaid.nl
http://www.cordaid.nl

Cottonwood Foundation

The Foundation is dedicated to promoting empowerment of people, protection of the environment, and respect for cultural diversity. The foundation focuses its funding on committed, grass roots organizations that rely strongly on volunteer efforts and where foundation support will make a significant difference. Cottonwood supports initiatives in the fields of reforestation (seedlings and tree-planting), sustainable soil fertility management practices, alternative energy and fuel wood resources etc. Eligible projects should meet all four of the following criteria: protect the environment; promote cultural diversity; empower people to meet their basic needs; rely on volunteer efforts.

Contacts: *Executive Director, Box 10803, MN 55110 White Bear Lake - United States, Phone: +1-651-4268797, Fax: +1-651-4260320*
info@cottonwoodfdn.org
http://www.cottonwoodfdn.org

Critical Ecosystem Partnership Fund

Critical Ecosystem Partnership Fund (CEPF). A joint initiative between Conservation International, The Global Environment Facility, the MacArthur Foundation and the World Bank. Investments support such projects as managing of protected areas and coordinating biodiversity corridors; training; trans-boundary planning; encouraging local dialogue with extractive industries; engaging in conflict resolution; priority setting and consensus building; strengthening indigenous organizations and facilitating partnerships between the private sector and protected areas. There are no specific deadlines for applications; however, CEPF will stop accepting proposals for a particular area or strategic direction once the designated funds are fully committed.

Contacts: *Critical Ecosystem Partnership Fund; Conservation International; 2011 Crystal Dr., Suite 500; Arlington, VA 22202 USA*
Phone: (1) 703 341-2400; Fax: (1) 703 553-0721
cepfgrants@conservation.org; http://www.cepf.net

Cultural Survival

CS provides empowerment tools that support indigenous efforts to effectively resist cultural degradation and threats to their habitat and sovereignty. CS is currently sponsoring projects initiated by indigenous peoples who seek to obtain land rights, protect and manage their natural resources, market sustainable products, preserve their language and art forms, etc. Furthermore Cultural Survival provides organizational support, administrative support, and acts as a fiscal sponsor to small, independent initiatives that contribute to Cultural Survival's mission.

Contacts: *215 Prospect Street, MA 02138 Cambridge - United States*
Phone: +1-617-441-5400; Fax: +1-617-441.5417
csinc@cs.org; http://www.culturalsurvival.org

DAMIEN FOUNDATION

Damien Foundation is a relatively small family foundation. Its aim is to create healthy grassroots environmental organizations and a high degree of co-operation and partnership among them. The Damien Foundation has a Small Grants Program to provide capacity building support and organizational assistance to small environmental organizations. It gives priority to organizations from Brazil.

Contacts: *235 Montgomery St., Ste. 1120, CA 94104 San Francisco – USA*
Phone: +1-415-421-7555; Fax: +1-415-421-0712.
http://www.damienfoundation.org

DFID: THE AFRICA ENTERPRISE CHALLENGE FUND

The Commission for Africa and G8 discussions highlighted that private enterprises are prime engines of growth and development. To boost economic growth in the region, a US$100 million multi-donor Africa Enterprise Challenge Fund (AECF) was recommended. The fund offers grants on a competitive basis to private companies to test innovative but unproven business models resulting in the greater participation of the poor in markets – both as consumers and also as workers and producers. The AECF is expected to stimulate US$200 million additional private sector investment in financial and agricultural markets across Africa.

Contacts: *http://www.africaenterprisechallengefund.org*

DFID: BUSINESS SECTOR ADVOCACY CHALLENGE FUND

The Business Sector Advocacy Challenge (BUSAC) Fund is a challenge fund intended to support private sector advocacy. Funds are provided, on a competitive basis, to support applications which meets BUSAC Fund's criteria and objectives and which are regarded as having the best chance of making positive impact on the sector. Applicants are responsible for the implementation of their actions, but could "buy in" appropriate support as they see fit.

Contacts: *http://www.busac.org/index.php*

DFID: CIVIL SOCIETY CHALLENGE FUND

The CSCF supports activities that: *(i)* improve the capacity of Southern civil society to engage in the local decision-making processes; *(ii)* improve the capacity of civil society to engage in national decision making processes; *(iii)* improve national linkages through global advocacy; *(iv)* provide innovative service delivery, and *(v)* provide service delivery in difficult environments. UK-based, non-profit organizations are eligible.

Contacts: *http://www.dfid.gov.uk*

Development Workshop Austria

EWA has specialized in regional development programs in the Sahel. The projects are planned and carried out in cooperation with grass-root organizations, farmers and artisan cooperatives, and other NGOs. The main focus of EWA's work lies on sustainable use of natural resources, conservation of the ecological equilibrium, diversification of income generation and the promotion of "rural artisans". In the context of desertification the sustainable use of water, soil and vegetation is of particular importance and therefore the EWA carries out the following programs: support, development and promotion of ecologically sound land-use systems with focus on agro-forestry, designed and implemented in cooperation with the affected population; promotion of sustainable water management, in particular well rehabilitation erosion protection, etc.; and support of rural infrastructure.

Contacts: *Thunstrasse 16, 5400 Hallein – Austria*
Phone: +43-662-627.112, Fax: +43-662-624.812,
EWA@magnet.at
http://www.ewa.or.at

DOC/EDA

The Economic Development Administration (EDA) of the U.S. Department of Commerce (DOC) announces policies and application procedures for its grants programmes to support projects designed to alleviate conditions of substantial and persistent unemployment and underemployment in economically-distressed areas and regions of the Nation, to address economic dislocations resulting from sudden and severe job losses and to administer the Agency's programmes.

Contacts: *Director, Grants and Contracts Division; Office of Science, SC-64 U.S, Department of Energy; 19901 Germantown Road; Germantown, MD 20874-1290*
http://www.doe.gov

Doen Foundation

Doen Foundation was established by the Dutch Postcode Lottery in 1991 to work towards the achievement of a livable world in which everyone has a place. Acting within its four fields of operation of sustainable development, culture, welfare and social cohesion, it provides subsidies where necessary, and arranges loans and equity investments where possible. However, DOEN only finances initiatives abroad on the area of Sustainable Development. DOEN's primary focus in the area of Sustainable Development consists of social-cultural aspects ('people'), economic aspects ('profit') and ecological aspects ('planet'). The application should give insight into the market and demand for products and the financial aspects of the proposed project (both budget and funding plan are realistic). In addition, the application must come from a legal body and have a broad national and international appeal, and be unable to obtain adequate financing locally. The applicant organization strives to become self-supporting and focuses on stimulating entrepreneurship, preferably sustainable entrepreneurship.

Contacts: *P.O. Box DOEN Foundation; Postbus 75621, 1070 AP Amsterdam, Netherlands;*
doen@doen.nl
http://www.doen.nl

Double Harvest

Double Harvest is a small family foundation and, although non-denominational; it works closely with Christian churches local to the project areas. The organization supports agricultural and reforestation projects. It provides land, materials, equipment, and technical instruction to establish nursery and farm areas used as teaching tools. Also projects in the field of crop and livestock development are supported. Double Harvest is active in several countries in Africa and Asia, as well in Haiti. The Dutch branch of Double Harvest is foremost active in Ethiopia.

Contacts: *http://www.doubleharvest.org*

Dow Chemical Company

Dow Chemical Company donates more than $18 million each year, globally. Funding criteria include the following: *(i)* address a need in a community where the company has a presence; *(ii)* provide hands-on science experiences for students below the college level; *(iii)* involve science, engineering or business at university level; or *(iv)* enhance the environment.

Contacts: *Building 47, Midland, Michigan 48667, Director–Global Contributions*
Fax: 517.638.7238
http://www.dow.com/about/corp/social/wwsi.htm#one

Drylands Coordination Group

The Drylands Coordination Group is a cooperation forum of five Norwegian NGOs involved in development cooperation in dryland areas in Africa: ADRA Norway, CARE Norway, Norwegian Church Aid, Norwegian People's Aid, and The Development Fund (Noragric, the Department for International Environment and Development Studies at the University of Life Sciences). DCG's vision is to improve the livelihood security of vulnerable households in drought-prone and marginal areas, especially in Africa. The group's objectives are: increased food security; sustainable management and use of natural resources; increased access to community based services such as health, education and clean water; competence building and institutional strengthening at local and regional levels. The group's activities can be divided into the following five categories: technical assistance to individual projects; identifying topics of common interest (e.g. through seminars and workshops); research and study assignments; information dissemination and networking; coordination and technical backstopping.

Contacts: *c/o Noragric P.O. Box 5001, 1432 Aas – Norway*
Phone: +47-64-949.823; Fax: +47-64-940.760
http://www.drylands-group.org

Dubai International Award for Best Practices to Improve the Living Environment

The award recognizes and enhances awareness of outstanding and sustainable achievements in improving the living environment as per the criteria established by the Second United Nations Conference on Human Settlements (Habitat II) and the Dubai Declaration. Submissions will be accepted under any of the following:

i. shelter and urban infrastructure;
ii. sustainable human settlement development;
iii. mainstreaming gender and social inclusion;
iv. improved consumption/production cycles;
v. urban poverty reduction and job creation;
vi. safe water supply and sanitation;
vii. Improved urban environment health
viii. Natural and human-made disasters;
ix. Responsiveness to the ideas and needs of youth;
x. Waste collection, recycling and reuse;
xi. Accessible public transport and communication;
xii. Crime prevention and social justice;
xiii. Efficient, accountable and transparent governance;
xiv. Use of information in decision-making;
xv. Architecture and urban design;
xvi. Addressing the needs of older persons;
xvii. Experimental and innovative practices.

Contacts: *Dubai Municipality; P.O. Box 67; Dubai, U.A.E;*
Phone: (971 4) 2215555; Fax: (971 4) 2246666;
info@dm.gov.ae
http://www.dm.gov.ae

Earth Love Fund

Earth Love Fund raises funds and awareness to support environmental conservation projects in tropical rainforests and endangered forests worldwide. Through this work, ELF also helps to empower and support indigenous communities, aiding their sustainable development and traditional ways of life. The fund supports projects with these goals:

protecting forests; developing a living from the forest as an alternative to forest clearances; supporting indigenous peoples; planting trees and crops; agricultural certification; ethno-biology; and raising awareness and building local expertise. The emphasis of this fund is on community-based conservation projects in rainforest regions and endangered forests world-wide. Funding is small to medium. Projects which promote and conserve the protection of endangered forests, biological diversity, and cultural and community integrity are given priority.

Contacts: *9 Standingford House Cave Street, OX4 1BA, Oxford - United Kingdom*
Phone: +44-1865-200208; Fax: +44-1865-209091
earthlove@gn.apc.org
http://www.unisong.com/elf/index.html.

Earth Preservation Fund

The nonprofit EPF was founded by the travel organization JOURNEYS. This organization has accepted a responsibility to preserve the resources we enjoy and over the years has supported numerous projects and ongoing programmes. EPF was the first wholly traveler-funded conservation organization in the United States. The Fund and JOURNEYS collaborate to create the highest form of participatory travel. Local community leadership and workmanship back contributions in the form of funds, labor or publicity on-site, which assure that goals are achieved. The EPF works to preserve local cultures and natural habitats around the globe. The Fund especially supports international community-based projects focused on environmental conservation and cultural preservation.

Contacts: *PO Box 7545, MI 48107 Ann Arbor - United States;*
Fax: +1-734-665.2945
info@earthpreservation.org
http://www.earthpreservation.org

EarthLink

EarthLink, the People and Nature Network, is a small German NGO. EarthLink together with its partners is involved in concrete projects to protect forests, oceans and coastal areas. Through environmental education for children and youth as well as adults and current decision makers from the political and business sector, EarthLink lays the foundation for ecologically sustainable development. Through forest-camps, joint reforestation programmes, seminars and workshops practically orientated ways for a future that is worth living in are shown.

Contacts: *Frohschammerstr.14, 80807 München – Germany.*
Phone: +49-89-3565.2102; Fax: +49-89-3565.2106
info@earthlink.de
http://www.earthlink.de

EcoVitality

This is a non-profit conservation planning and eco-marketing NGO set up to design, fund, and operate integrated conservation and development (ICAD) programmes protecting ecosystems and wildlife of developing countries. EcoVitality do not have the financial resources to fund projects at present. Instead, they raise funds to create conservation-compatible businesses in ecologically vulnerable areas, market the goods or services in wealthy developed states, and send the profits back to the producing communities in return for better conservation outcomes. This approach links new economic opportunities directly to stronger environmental commitments. EcoVitality is also involved in improving the competitive position of marine and forest products produced in an environmentally sustainable manner.

Contacts: *224 Centre Street, 2d Floor, NY 10013, New York – USA,*
Phone: +1-212-966.8803; Fax: +1-212-966.8803
ecovitality@ecovitality.org
http://www.ecovitality.org

Eirene International

This is an ecumenical international peace and development service organization. It was founded in 1957 by Christians from several denominations and apart from its head office in Germany; it has branch offices in The Netherlands and Switzerland. They focus their work on promoting intercultural dialogue, peace, community development, environment, human rights, and education.

Contacts: *Postfach 1322, 56503 Neuwied –Germany*
Phone: +49-2631-83790; Fax: +49-2631-31160
eirene-int@eirene.org
http://www.eirene.org

Entwicklungswerkstatt Austria

The name means Development Workshop Austria. The overall objective of EWA is to support of self-help activities through the provision and transfer of know-how and methods for personal and vocational development and the support of project partners and the local population in order to develop financial instruments and management capacities. EWA has specialized in regional development programmes in the Sahel in West Africa. The main focus of EWA's work lies on sustainable use of natural resources, conservation of the ecological equilibrium, diversification of income generation and the promotion of "rural artisans". EWA is currently involved in the implementation of a total of 10 projects in Burkina Faso and in Senegal.

Contacts: *Thunstrasse 16, 5400 Hallein –Austria*
Phone: +43-662-627.112; Fax: +43-662-624.812
EWA@magnet.at; http://www.ewa.or.at

European Union and Commission Programmes

i) *Promotion of the conservation and sustainable management of tropical forests and other forests in developing countries:* The general aim is to promote the conservation and sustainable management of tropical forests and other forests in developing countries, so as to meet the economic, social and environmental demands placed on forests at local, national and global levels. Activities under this programme aim at: Raising the status of forests in national policies and integrating forest policies based on sustainable forest management in development planning; promoting the production and use of wood and non-wood forest products from sustainably managed resources; contributing to the adequate valuation of forest resources and services; ensuring active participation of forest-dependent people and local communities in the development of national forest policies and in development planning; improving coordination and the flow of information between Commission and Member State projects so as to put in place coherent actions in that area. *EuropeAid Co-operation Office Unit F4, Office: J-54 6/41, 1049 Brussels.*

ii) *EU tropical forest research network:* The ETFRN is a forum for communication between European organizations, researchers, EU institutions and others concerned with (sub-) tropical forest research. *http://www.etfrn.org*

iii) *Seventh framework programme for research and technological development:* The European Commission (EC) is providing 20 million euros to allow scientists in developing countries to join existing European projects. The funds are intended to address poor participation by 'third countries' in projects funded by the EU's Seventh Framework Programme for Research and Technological Development. The EC invites young or experienced scientists from both the public and private sector to join ongoing projects in priority areas, such as biotechnology for health. *http://cordis.europa.eu.*

Face Foundation

The Face Foundation was set up to help abate the enhanced greenhouse effect by planting and protecting forest. Its slogan is: More Forest, Less CO2. Face works with partners who are genuinely interested in forests and are able to manage these forests sustainably. An independent organization certifies the forests and verifies the amounts of CO2 they store.

Contacts: *Postbus 646, 6800 AP Arnhem - the Netherlands*

Phone: +31-26-357.0770; Fax: +31-26-357.0777

face@facefoundation.nl

http://www.facefoundation.nl

Farm-Africa

FARM-Africa aims to reduce poverty by enabling marginal African farmers and herders to make sustainable improvements to their well-being through more effective management of their renewable natural resources. FARM works in partnership with communities, governments, local organizations, international NGOs and the private sector to develop strong rural livelihoods and then share the results with others for it to have maximum impact. FARM works with a wide range of small-scale farmers and herders through its country programmes. The focus is on developing innovative approaches to natural resource management and communicating good practice to policy makers, government and NGOs.

Contacts: *Chief Executive, 9-10 Southampton Place, London WC1A 2EA – United Kingdom*
Phone: +44-20-7430.0440; Fax: +44-20-7430.0460
farmafrica@farmafrica.org.uk
http://www.farmafrica.org.uk

Florence and John Schumann Foundation

The foundation focuses its support on projects that attempt to invigorate democracy, with special consideration given to projects that encourage greater citizen awareness of the influence of private money on the conduct of the public's business. Grants are made to organizations, including grassroots activists, whose goals are to increase the accountability of government and corporate decisions and to democratize the financing of elections. There is no standard application form.

Contacts: *07042 USA NJ Montclair 33 Park Street*
Phone: (201) 783-6660; Fax: 201-783-7553
http://www.undueinfluence.com/schumann_foundation.htm

Floresta Inc.

This Christian non-profit organization, works to reverse deforestation and poverty in the world by transforming the lives of the rural poor. Floresta's mission of transformational development occurs through planting, teaching, learning, creating enterprise, and sharing the Good News of the Gospel of Jesus Christ. It offers technically appropriate, business based programs that lead to self-sufficiency and sustainability.

The program uses sound business-based economic principles to motivate farmers to change their practices. It provides medium and short term loans through their agro-forestry revolving loan fund. When products are ready, it also offers assistance in taking advantage of the local and global marketplace.

Contacts: *4903 Morena Boulevard, Suite 1215, CA 92117 San Diego – USA*
Phone: +1-800-633.5319; Fax: +1-619-274-3728
floresta@xc.org
http://www.floresta.org

Forest Management Foundation

The FMF was set up as a development organization to promote community-based forestry. FMF believes that, for forestry projects to become viable and sustainable in the broadest sense, the environmental, social, and economic aspects of these projects must be properly integrated. To assist in this process, FMF provides a range of services focused in particular on the economic and commercial side. FMF works in partnership with people who plan and manage community forestry projects in order to help them to run projects in an efficient, economic and effective way. FMF can assists in drawing up project proposals and finding possible sources of funding for well-conceived community forestry projects. The foundation is active in Africa, the Pacific and Ecuador.

Contacts: *6 Salisbury Avenue, Penarth, CF64 3JA Cardiff - United Kingdom*
Phone: +44-1222-70.74.44; Fax: +44-1222-71.11.33
fmf@fmf.u-net.com
http://www.btinternet.com/~miombo.forum/forest.html

Forest Peoples Project

The Project was established in 1999 as the charitable arm of the Forest Peoples Programme to provide support for forest peoples, particularly in sustainable development, education and capacity building. Its aims are to support indigenous and tribal forest peoples to: promote their collective and individual rights; secure their lands and manage their natural resources; carry out sustainable community development. FPP works to achieve its aims by means of: technical support, capacity building and policy advice for forest peoples; networking with NGOs, indigenous support organizations and agencies; researching and analyzing the situation of forest peoples; and providing information on forest peoples. Its funds are limited.

Contacts: *1c Fosseway Centre, Stratford Road, GL56 9NQ,*
Moreton-in-Marsh - United Kingdom
Phone: +44-1608-652893; Fax: +44-1608-652878
fpproject@gn.apc.org;
http://www.forestpeoples.gn.apc.org

Foundation Ensemble

The projects supported by Foundation Ensemble must come within the scope of its mission: provide target populations with significant well-being, taking into account the link between people and nature. The projects given precedence are those which meet identified needs. They foster the emergence and/or reinforcement of the partners' capacities.

The parties involved in the project must be clearly identified by nature and by socio-economic group, and the solutions proposed should be adapted to each category. The project must imply the equitable representation of the various interests by means of participation. Organizations wishing to submit a project are invited to send a three-pages presentation.

Contacts*: Fondation Ensemble; 45 rue de Babylone; 75007 Paris France*
info@fondationensemble.org; http://www.fondationensemble.org

FRIENDS OF CONSERVATION

FOC is a conservation charity that works with local communities across the world to protect their surrounding wildlife and habitats. Its mission statement is to join with local people and partners to support sustainable ways of living in harmony with wildlife, ecosystems and the environment. FOC is committed to working with local peoples to develop and promote a balance between their needs and those of the wildlife with which they share their natural habitat and ecosystems. Projects aim at protection of wildlife (including research), conservation of habitats (including environmental education and reforestation programmes), and community development. FOC's Travel and Tourism Conservation Committee (TTCC) aims to increase co-operation between the tourism industry and communities to preserve local ecosystems. The key objective of the TTCC is: raising funds for FOC UK, to support global conservation, community and education projects and promote those objectives within the industry to tourists.

Contacts: *UK Office: 16-18 Denbigh Street, SW1V 2 ER London - United Kingdom*
Phone: +44-20-7592.0110; Fax: +44-20-7828.4856
focinfo@aol.com
http://www.foc-uk.com

USA Office: 1520 Kensington Road, Suite 212, IL 60523-2156 Oak Brook – US
Phone: +1-630-954.3388; Fax: +1-630-954.1016
info@friendsofconservation.org

Kenya office: Left Wing, Great Jubilee Centre, Langata Road, Nairobi – Kenya
Phone: +254-20-890143; Fax: +254-20-890147
info@fockenya.org

FUNDACIO NATURA

Fundació Naturais one of the first NGO created in Spain exclusively devoted to the conservation of the biological diversity through the protection, improvement and restoration of the natural ecosystems. Fundació Natura achieves its objectives through the purchase, leasing, transference of land property, management agreements with landowners or the purchase of resource exploitation rights. Thus, a network of protected natural areas is created and managed to conserve and improve its biodiversity. It also encourages the restoration of highly natural valued areas, which have suffered a process of degradation (due to fires or other human alterations) so that they will regain their original natural value. Whenever possible, it tries to reconcile public use and conservation.

Contacts: *Rambla de Catalunya, 121, 6è 9a, 08008 Barcelona – Spain*
Phone: +34-93-237.01.81; Fax: +34-93-237.01.81

info@fundacionatura.org
http://www.fundacionatura.org

FUNDESO

Fundeso's primary attention is directed towards socio-economic programmes centered on reinforcement of the economical position of small producers and the professional education of young people and adults. Fundeso promotes socio-economical development programmes in developing countries, with special attention for Latin America. Themes are amongst others forest issues like afforestation and non-timber forest products.

Contacts: *Gran Vía, 16-4 izq, 28013 Madrid – Spain*
Phone: +34-91-7014700; Fax: +34-91-7014701
fundeso@fundeso.org
http://www.fundeso.org

GEO SCHÜTZT DEN REGENWALD

Trees for Life" was established in 1990 by the magazine GEO. The aim is to protect tropical (rain) forests and to improve the economic situation of its inhabitants and neighbours. Funding preference is for agro-forestry projects so as to develop sustainable ways of using the rainforest. They fund infrastructural measures which serve to protect selected areas; technical support, health care, and educational aid for indigenous peoples; information dissemination of the environmental situation in Southern countries.

Contacts: *c/o Redaktion GEO: Am Baumwall 11, 20444 Hamburg – Germany*
Phone: +49-40-3703-0; Fax: +49-40-3703-5648
http://www.geo.de/projekte/regenwald

GERMAN AGRO ACTION

Welthungerhilfe (German Agro Action) is one of the largest non-governmental development organizations in Germany. GGA fights poverty in developing countries in order to improve living conditions and secure a dignified existence. One of the focus areas of GGA's project work concentrates on agriculture and the environment. The spectrum of projects ranges from reforestation measures, erosion prevention and building wells to special programmes supporting women, children and youth.

Contacts: *Postfach 120509, 53000 Bonn – Germany;*
Phone: +49-228-2288.0; Fax: +49-228-220710
info@welthungerhilfe.de
http://www.welthungerhilfe.de

Global Development Network

The Global Development Awards and Medals Competition is the largest international contest for research on development. It has supported multidisciplinary research on a range of issues including global health concerns and domestic responses; pro-poor market reform; changes in global trade; industrial development and long-term growth; governance and development, reforms, interest groups and civil society; conflict, human security and migration; and the role of institutions for development in the context of globalization.

Contacts: *Global Development Network; 2nd Floor, West Wing, ISID Complex; Plot No. 4; Vasant Kunj Institutional Area; New Delhi – 110070, India*
Phone: + (91) 11-2613-9494 / 2613-6885; Fax: + (91) 11-4170-4248 / 2613-6893; US
http://www.gdnet.org

Global Environment Facility

GEF is the designated financial mechanism for international agreements on biodiversity, climate change, and persistent organic pollutants. The GEF funds projects in six focal areas: biodiversity, climate change, international waters, ozone, land degradation (desertification), and persistent organic pollutants. Taking into account the objectives of the Convention to Combat Desertification (CCD), dozens of GEF projects cut across the focal areas to address land degradation. Through the Small Grants Programme (SGF), GEF aims to deliver global environmental benefits in the GEF Focal Areas of biodiversity conservation, climate change mitigation, protection of international waters, prevention of land degradation (primarily desertification and deforestation), and elimination of persistent organic pollutants through community-based approaches. All grants are coordinated by the National Coordinator for the country. A list of SGP eligible countries are listed on the website, along with country-specific contact information.

Contacts: *1818 H Street, NW, MSN G 6-602, DC 20433 Washington - United States*
Phone: +1-202-473.0508; Fax: +1-202-522.3240/3245
gef@gefweb.org
http://www.gefweb.org
http://www.sgp.undp.org

Global Finland

The Department for International Development Co-operation, formerly known under the name FINNIDA, is the department in the Ministry of Foreign Affairs of Finland responsible for international development co-operation. One of the long-term development objectives of the Finnish development cooperation is combating global and local environmental threats. The International Environment Policy Unit is responsible for International NGOs, which concentrate their work on international and regional environmental issues such as climate policy (mitigation and adaptation), water and sanitation, sustainable forest management, watershed management and land degradation (e.g. desertification), biodiversity as well as implementation of sustainable development. Funding is primarily directed towards the international environmental activities where the work by the International NGOs is an extension of the work Finland is already engaged in. The main focus is to assist cooperation with developing countries, especially in developing their capacity building and awareness on environmental issues.

Contacts: *International Environment Policy Unit, P.O. Box 176, 00161 Helsinki – Finland*
Phone: +358-9-16056417; Fax: +358-9-1605.6375
http://www.global.finland.fi

GLOBAL FUND FOR WOMEN

The Global Fund for Women is an international network of women and men committed to a world of equality and social justice, advocates and defenders of women's human rights by making grants to support women's groups around the world. The Global Fund makes grants to seed, strengthen and link women's rights groups based outside the United States working to address human rights issues. The Global Fund receives over 3,000 proposals a year and is able to award about 600 grants. Applications are accepted throughout the year and grants awarded every three months.

Contacts: *Main Office, Global Fund for Women; 1375 Sutter Street, Suite 400; San Francisco, CA 94109 USA*
Phone: (415)202-7640; Fax: (415)202-8604
http://www.globalfundforwomen.org

GLOBAL GREENGRANTS FUND

The mission of the Fund is to help protect the global environment by strengthening the grassroots environmental movement in developing countries with small grants. The GGF supports a wide range of community-based initiatives that protect the environment in the most underserved and threatened regions on Earth. The purpose of GGF is to strengthen organizational infrastructure and networking capability with grants for those essential expenses that allow organizations to do their work.

Contacts: *2840 Wilderness Place Suite E, CO 80301 Boulder - United States*
Phone: +1-303-939.9866; Fax: +1-303-939.9867
info@greengrants.org
http://www.greengrants.org

GLOBAL2000

Global2000 is the largest and most well-known Austrian environmental protection organization. It coordinates and administrates rainforest projects to support the indigenous population in Brazil and Columbia. The objectives of the projects are to support indigenous cultures in the actualization of self-determined and careful development strategies, the long-term maintenance of their habitats and a rash counteraction to the destruction of the rainforests.

Contacts: *Flurschützstraße 13, 1120 Vienna – Austria*
Phone: +43-1-812 57 30-0; Fax: +43-1-812 57 28
office@global2000.at
http://www.global2000.at

Gorta

Gorta was established in October 1965 and is Ireland's first nondenominational Third World development agency, specifically set up to prevent famine by established self-sufficiency community-controlled projects. Gorta concentrates its development activities in Sub Saharan Africa, but has continued to fund existing partners on other continents.

Contacts: *12 Herbert Street; Dublin 2 Ireland*
Phone: + 353 1 661 5522, Fax: + 353 1 661 2627
http://www.gorta.org

Green Development Foundation

GDF seeks to contribute structurally to poverty alleviation and to increase the possibilities and opportunities of development for marginalized social groups. The Foundation does not fund projects. Instead, they assist and advise grassroots organizations in formulating, implementing and evaluating development projects that are socially equitable, ecologically sound and economically sustainable.

Contacts: *Leliegracht 19, 1016 GR Amsterdam - The Netherlands*
Phone: +31-20-421.80.98; Fax: +31-20-421.80.13
http://www.antenna.nl/greendevelopment

Heinz Endowments

The Endowments supports projects designed to improve the quality of life and address challenges it shares with communities across the United States. The focus is on five disciplines represented by grant-making programme areas: Arts and Culture; Children, Youth and Families; Education; Environment; and Innovation Economy.

Contacts: *http://www.heinz.org*

Helvetas

This organization works with partner NGOs in over 20 countries in Africa, Asia and Latin America promoting self-reliant and sustainable development with an ecological awareness and the interests of future generations. In the area of ecology and forest conservation, Helvetas concentrates its work in the following fields: land use planning, soil conservation, afforestation, biodiversity, rural credit systems, promotion of peasant organizations, marketing of forest and agricultural products.

Contacts: *St. Moritzstr. 15, 8042 Zurich – Switzerland*
Phone: +41-1-368.65.00; Fax: +41-1-368.65.80
info@helvetas.ch
http://www.helvetas.ch

HUMANE EARTH FOUNDATION

The objectives of the Foundation are to: defend the natural environment and biodiversity; combat poverty and exclusion, especially concerning food safety and wholesomeness, together with access to quality health care. The foundation supports especially projects that: (i) Inform people of the risks that they face because of attacks on the environment and biodiversity; (ii) implement concrete experience leading to changes in behaviour in a way that better respects the environment; (iii) spread knowledge of successful experience, to show that change is possible.

Contacts: *Fondation pour une Terre Humaine; 15 route de Fribourg;*
1723 Marly2 Switzerland
Phone: + 41 26 435 33 70; Fax: + 41 26 435 33 71
infos@terrehumaine.org
http://www.terrehumaine.org

HUMANIST INSTITUTE FOR CO-OPERATION WITH DEVELOPING COUNTRIES

HIVOS seeks to help improve the opportunities and scope for development of people in the South. Respect for human rights, pluralism and democratization are essential elements here. HIVOS has an integrated approach towards environment and development issues. With respect to environment issues, focus is on sustainable production/consumption and related management of land and (agro) biodiversity. Hivos supports environmental organizations and action groups that are in the vanguard of their own society at the interface between the environment and development.

Contacts: *Raamweg 16, 2596 HL Den Haag – The Netherlands*
Phone: +31-70-376.55.00; Fax: +31-70-362.46.00
hivos@hivos.nl; http://www.hivos.nl

IFAD/NGO EXTENDED COOPERATION PROGRAMME

The International Fund for Agricultural Development (IFAD) was created to mobilize resources on concessional terms for programmes that alleviate rural poverty and improve nutrition. Unlike other international financial institutions, which have a broad range of objectives, the Fund has a very specific mandate: to combat hunger and rural poverty in developing countries. The purpose of IFAD/NGO ECP is to enhance IFAD's direct collaboration with NGOs in the promotion of participatory and community-based rural development and poverty alleviation.

Contacts: *Coordinator NGO and Civil-Society Partnerships,*
c/o IFAD Via del Serafi co 107, 00142 Rome – Italy
Phone: +39-6-5459.2238; Fax: +39-6-5459.2238
http://www.ifad.org/ngo

Institut Pan Africain Pour le Developpement

Training courses lasting up to four weeks each in six areas: Management and Funding for Development, Gender and Development, Agriculture and Rural Development, Politics and Education Systems, Development of small and medium-sized enterprises, Environment, Health and Community Development.

Contacts: *01 BP 1756 Ouagadougou, Burkina Faso,*
Tel: (226) 36 48 07 or (226) 36 47 62, Fax :(226) 36 47 30
ipd_aos@cenatrin.bf

Interchurch Organization for Development Co-operation

ICCO finances activities that address poverty and injustice. The main focal points of ICCO are human rights, gender and environment, social development, and organizational and institutional development. With regard to the environment, ICCO funds projects on sustainable forestry and sustainable land use.

Contacts: *Postbus 151, 3700 AD Zeist - the Netherlands;*
Phone: +31-30-692 78 11; Fax: +31-30-692 56 14
http://www.icco.nl

Intercooperation

The aim of Intercooperation is to alleviate poverty through rural development. The focus is on improving rural livelihoods through the sustainable management, harvesting and processing of natural resources in order to promote ecologically, economically and socially sustainable development. Intercooperation's expertise, process competence and services are grouped around three principal working domains: natural resource management aiming at a balance between production and protection for the benefit of the poor - within this domain Intercooperation works on specific themes, ranging from agricultural production and tropical forestry to climate change and biodiversity, watershed programmes; rural economy issues especially the availability of infrastructure and the existence and accessibility of basic social services; and local governance and civil society in order to promote the development of municipalities, villages and other local bodies.

Contacts: *PO Box 6724, 3001 Bern – Switzerland;*
Phone: +41-31-382.10.10; Fax: +41-31-382.10.09
http://www.intercooperation.ch

INTERNATIONAL TREE FOUNDATION

The ITF was founded in 1924 by Dr. Richard St Barbe Baker, one of the pioneers of tree planting in Britain and the world. On a global scale, ITF tree planting grants help the world's poorest environments and societies. ITF works towards a world richer in trees, by planting, preserving and educating people on the value of trees worldwide. Since 2002 the Foundation has focused its grants on specific areas of Africa and India to ensure that projects get sufficient funding and are followed through properly. ITF is involved in the establishment and maintenance of tree nurseries, tree planting and agro-forestry projects.

Contacts: *Sandy Lane, RH10 4HS Crawley Down, West Sussex - United Kingdom*
Phone: +44-1342-712.536; Fax: +44-1342-718.282
http://www.internationaltreefoundation.org

IRISH AID CIVIL SOCIETY FUND

Irish Aid, the Government's official development assistance programme, invites applications under the Civil Society Fund (CSF) from Irish civil society organizations. The CSF supports non-governmental organizations and other civil society organizations to respond to the development needs of poor communities overseas. Irish Aid's overall policy goals are poverty reduction and sustainable, equitable development.

Contacts: *Irish Aid; Department of Foreign Affairs; Bishops Square,*
Redmond Hill; Dublin 2 Ireland
Phone: 53 1 408 2000, Fax: 353 1 408 2880
http://www.dci.gov.ie

JAMES S. MCDONNELL FOUNDATION: 21ST CENTURY SCIENCE INITIATIVE

In 2000 the James S. McDonnell Foundation (JSMF) announced new grant making guidelines for a new funding initiative: the 21st Century Science Initiative. The JSMF believes that private philanthropic support for science is most effective when it invests in the acquisition of new knowledge and in the responsible application of knowledge for solving the real world problems. Applicants are encouraged to keep this in mind when preparing proposals. Projects supported through the 21st Century Science Initiative are expected to meet highly selective intellectual standards. One focus is the Complex Systems programme, which supports scholarship and research directed toward the development of theoretical and mathematical tools that can be applied to the study of complex, nonlinear systems. It is anticipated that research funded in this programme will address issues in fields such as biology, biodiversity, climate, demography, epidemiology, technological change, economic development, governance, or computation. Proposals attempting to apply tools and models to problems where such approaches are not yet considered usual or mainstream (for example, differentiating normal physiology from disease) are encouraged.

Contacts: *The James S. McDonnell Foundation; 1034 South Brentwood Blvd., Suite 1850; Saint Louis, MO 63117 USA*
Phone: (314) 721-1532; Fax: (314) 721-7421
http://www.jsmf.org

JAPAN FUND FOR GLOBAL ENVIRONMENT

JFGE is part of the Environmental Restoration and Conservation Agency of the Japan Environment Corporation. Its aim is to provide support for NGO programs working for the conservation of the environment. JFGE contributes through two programs: the Grants Program, which provides financial support for environmental conservation projects conducted by NGOs; and the NGO Support Program, which provides support to NGOs through provision of information materials and training courses (only in Japanese and for Japanese-speaking). Projects eligible for assistance are: actual implementation of afforestation, desertification, nature conservation, wildlife protection and pollution control (air, water and soil) projects; field studies and research directly related to the above projects; projects aimed at achieving sustainable management of natural resources (local and regional level); projects aimed at raising environmental awareness among the public; provision of skills training and information materials to both NGOs and local communities for the implementation of their environmental projects; organizing of conferences, seminars and workshops; and other conservation projects such as utilization of clean energy. JFGE requires all applications to be written in Japanese. Therefore, foreign NGOs must have a Japanese representative (resident in Japan), who can coordinate and fill out all the necessary form in Japanese.

Contacts: *1-4-1, Kasumigaseki, Chiyodaku, 100-0013 Tokyo – Japan*
Phone: +81-3-5251.1076; Fax: +81-3-3592.5090
http://www.jec.go.jp

JOHN AND MARY R. MARKLE FOUNDATION – POLICY FOR A NETWORKED SOCIETY

The Foundation rarely funds unsolicited proposals, but seeks partners with the capabilities to complement its projects and aid in its vision. Approved applications are entered into a database for further reference.

Contacts: *The Markle Foundation; 10 Rockefeller Plaza, 16th floor; New York, NY 10020*
proposals@markle.org
http://www.markle.org

JPMORGAN CHASE FOUNDATION

The JPMorgan Chase Foundation welcomes grant inquiries from non-governmental organizations working internationally in the Foundation's three focus areas: community development; youth education; and arts and culture. Of particular interest are requests in the areas of micro-finance and support for small business development; low-income housing; youth education in low-income communities; and arts and culture projects aimed at promoting asset development in low-income communities.

Contacts: *http://www.jpmorganchase.com*

KHWARIZMI INTERNATIONAL AWARD

The Khwarizmi International Award recognizes the efforts made by researchers, innovators and inventors both in Iran and all over the world and to appreciate their invaluable achievements and contributions to various fields of science and technology. The Award is held annually. The KIA is awarded to projects which belong to four main categories: *(i)* innovation; *(ii)* invention; *(iii)* applied Research; *(iv)* fundamental Research Projects must have made outstanding contributions to the following fields of Science and Technology: Engineering; Basic Sciences; Human sciences; Agriculture; Biotechnology; Emerging technologies.

Contacts: *Khwarizmi International Award; No 71, Shahid Mosavi St.; Enghelab Ave.*
P.O. Box 15815-3538; Tehran, Islamic Republic of Iran
Phone/ Fax: (+98 21) 88838341
khwarizmi@irost.org
http://www.khwarizmi.irost.ir

KING BAUDOUIN INTERNATIONAL DEVELOPMENT PRIZE

The aim of the King Baudouin International Development Prize is to acknowledge the work of persons or organizations which have made a substantial contribution to the development of countries in the southern hemisphere or to solidarity between industrialized nations and developing nations. It also seeks to remind public opinion that the problems of development are today even more acute than when they gradually emerged in the collective awareness of nations in the aftermath of the Second World War. The prize is awarded every other year by the Foundation's Board of Governors. Beyond its actual financial value, the Prize provides winners international visibility and publicity, with the main agents of development in particular, such as the United Nations and its specialized agencies, the World Bank, the European Union and a number of bilateral development agencies, the world of foundations or international NGOs.

Contacts: *The King Baudouin Fndn; 21 Rue Brederodestraat; B-1000 Brussels Belgium*
Phone: 32-2-549.02.73; Fax: 32-2-500.54.31
info@kbprize.org
http://www.kbprize.org

KUWAIT FUND

In December 1961, the Kuwait Fund for Arab Economic Development was established as the State of Kuwait's agency for the provision and administration of financial and technical assistance to the developing countries. The operations of the Fund were originally confined, in accordance with its initial mandate, to the Arab countries. In July 1974 the scope of the Fund's activity was extended to the rest of the developing world, and its capital was increased. Project proposals submitted for consideration by the Fund should have a strong public sector element and be of a developmental nature. The Fund is not subject to restrictions as to the sectors it may support and the project may fall in any of the sectors of agriculture and irrigation, transport and communications, energy, water supply, sewage, industry, education and health.

Contacts: *Kuwait Fund for Arab Economic Development*
Mirqab; Mubarak Al-Kabeer St.; Kuwait City; P.O. Box 2921 Safat 13030 Kuwait;
State of Kuwait
Phone: (+965) 2999000
info@kuwait-fund.org
http://www.kuwait-fund.org

LEVI STRAUSS FOUNDATION

The Foundation has identified three specific program areas for having the most significant impact to alleviate poverty among women and youth. These are building assets; preventing the spread of HIV/AIDS by addressing stigma and discrimination and funding education and harm-reduction programs such as syringe access; and improving the working and living standards of garment workers through a global workers' rights initiative.

Contacts: *Levi Strauss South Africa P/L, Safmarine House, 22 Riebeek St.*
4th Fl., Cape Town 8001, South Africa
Phone :(27).21.403.9400; Fax: (27).21.419.1713
http://www.levistrauss.com

LEVINSON FOUNDATION

The Levinson Foundation is a private family foundation that has been in existence since 1955 and makes grants to individuals and groups committed to developing a more humane and rewarding society, in which people have a greater ability and opportunity to determine directions for the future. Areas of interest include: The environment: including protection of ecosystems and biological diversity; alternative energy and protection of the atmosphere; alternative agriculture and transportation; breaking the link between illegal resource extraction, civil conflict and markets; and the development of the Israeli Environmental Movement.

Contacts: *http://www.levinsonfoundation.org/index.html*

LIVING EARTH

Living Earth involves a wide range of stakeholders in its work - people in corporations, communities, schools, and government institutions. The organization specializes in environmental learning and in developing tri-sector partnerships among corporations, government agencies, and civil society.

Contacts: *4 Great James Street, WC1N3DA, London -United Kingdom*
Phone: +44-20-7440.9750; Fax: +44-20-7242.3817
info@livingearth.org.uk
http://www.livingearth.org.uk

LUDWICK FAMILY FOUNDATION

Founded in 1990 by Arthur and Sarah Ludwick, the Ludwick Family Foundation is a philanthropic organization committed to assisting a broad array of groups that are working to make a positive difference in the world.

Contacts: *http://www.ludwick.org*

MAENDELEO AGRICULTURAL TECHNOLOGY FUND

MATF is offers competitive grants to provide opportunities to institutions involved in technology generation and/or dissemination to work in partnership to develop and apply innovative mechanisms for transferring proven agricultural technologies to farmers and other end-users. The key elements of the strategy adopted by the Fund include: facilitating linkages between diverse technology transfer practitioners, encouraging greater participation of farmers in technology transfer processes and using innovative cost-effective transfer approaches and methods in order to enhance the dissemination and use of existing viable technologies.

Contacts: *Studio House, 4th Fl.; Argwings Kodhek Road, Hurlingham*
The Fund Manager, MATF; PO Box 49502; 00100, Nairobi, Kenya
Phone: +254 20-2732044/2203/1664
info@maendeleo-atf.org
http://www.maendeleo-atf.org

MAZON: A JEWISH RESPONSE TO HUNGER

MAZON is a citizens' movement, the outgrowth of tens of thousands of American Jews who have stepped forward to solve one of the world's most devastating – and most preventable – problems. With their support, MAZON works around the United States, and around the globe, to bring critical relief to millions of hungry families. MAZON allocates donations from the Jewish community to prevent and alleviate hunger among people of all faiths and backgrounds. MAZON's international grant-making principally funds programs that strengthen poor families' ability to feed themselves, and that promote equitable practices that enable hungry people to access the food they need — whether through greater access to land, more control over the food production processes or a better social safety net. International grantees also include several direct feeding programs.

Contacts: *1990 South Bundy Drive, Suite 260; Los Angeles, CA 90025 USA*
Phone: (310) 442-0020; Fax: (310) 442-0030
http://www.mazon.org

McKnight Foundation: Africa Grants Programme

The McKnight Foundation supports programs that provide community and family services for those in need; preserve the environment; support the arts; and promote scientific research in selected areas. The Africa Grants Program's goal is to foster women's social and economic empowerment. The Africa program to supports efforts that: *(i)* foster women's ability to start, develop and manage their own small and micro businesses. Funded programs provide financial services as well as access to new skills or technologies that will improve productivity. *(ii)* enhance the ability of rural women to grow crops and raise livestock to achieve food security and increased income. Farm-based programs promote sustainable organic agriculture, building household assets through animal husbandry, and women's entry to commercial farming through high value crops and access to new markets; *(iii)* build women's ability to actively manage natural resources in order to conserve or increase income, or save time or labor. This category of programs supports business development around park land as an alternative to encroaching on protected areas; well-managed commercial responses to market demand for natural resources; and technologies that conserve both scarce resources and women's time.

Contacts: *The McKnight Foundation; 710 South 2nd St., Suite 400; Minneapolis, MN 55401 USA*
Phone: 612-333-4220; Fax 612-332-3833
http://www.mcknight.org/international

Mellemfolkeligt Samvirke

The Danish Association for International Co-operation combines development assistance with both political lobbyism and active grassroots. MS is working to promote understanding and solidarity across borders, colour and politics; and to contribute to sustainable global development and a just distribution of the earth's resources. MS strives to do that by means of development assistance, youth exchange, development education, ethnic minorities work, information and advocacy work. In its goal to fight poverty MS supports its partners in the South with different types of assistance such as training, transportation, small-scale funding, equipment, support for employment of qualified local people - and the posting of Danish development workers. Several country programmes concentrate on protection and conservation of the environment through sustainable use of local natural resources.

Contacts: *Borgergade 14, 1300 Copenhagen – Denmark*

Phone: +45-7-731.00.00; Fax: +45-7-731.01.01
http://www.ms.dk

MONSANTO FUND

The Monsanto Fund was established in 1964 and their philanthropic goal is to bridge the gap between people's needs and their available resources. The Foundation seeks to help people realize their dreams, and hopefully inspire them to enroll others in their vision. The focus of their funding falls into four Priority Areas: *(i)* improving nutritional well-being through agriculture, *(ii)* the environment, *(iii)* science education and, *(iv)* our communities.

Contacts: *Monsanto Nairobi; IandM Building, 6th Floor; 2nd Ngong Avenue, Ngong Road; Nairobi, Kenya*
http://www.monsantofund.org

NAGAO NATURAL ENVIRONMENT FOUNDATION

The Nagao Natural Environment Foundation (NEF) is a non-governmental organization dedicated to promoting nature conservation in developing countries. The NEF pursues its objectives through two main schemes: Research grant and scholarship. The research grant scheme provides grants to creative and committed local people, allowing them to undertake research and related activities in within the scope of nature and conservation. The scholarship scheme supports local students who can work for nature and wildlife conservation in the future. Applicants must be local residents in developing countries.

Contacts: *Nagao Natural Environment Foundation (NEF); 3-10-10, Shitaya, Taito-ku, Tokyo 110-0004, Japan*
Phone: +81-3-3824-0771; Fax: +81-3-3824-0772
http://www.jwrc.or.jp/NEF

NATURE AND CULTURE INTERNATIONAL

NCI is dedicated to preserve lands containing high biodiversity and great numbers of threatened and endangered species; to support traditional cultures which can be as threatened as their natural environments; to promote basic scientific research in biological and cultural diversity. Other themes are environmental education, sustainable development, and capacity building of partners.

Contacts: *1400 Maiden Lane, CA 92014 Del Mar - United States*
Phone: +1-858-259.0374; Fax: +1-858-259.1815
http://www.natureandculture.org

Nepenthes

Nepenthes was set up in 1983 as a non-profit organization by a group of biologists, anthropologists and others who wanted to work for preserving the tropical rain forests of the world and help their indigenous people. Its purpose is threefold: *(i)* to propagate the knowledge of tropical rainforests and other natural forests and the problems associated with their destruction; *(ii)* to work for a long-term sustainable development that will contribute to a rational utilization of forests and conservation of natural forests; and *(iii)* to support indigenous peoples in regions with natural forests in conserving their forests and in strengthen their cultural identity and right of self-determination.

Contacts: *PO Box 5102, 8100 Aarhus – Denmark*
Phone: +45-86-135232; Fax: +45-86-125149
http://www.nepenthes.dk

Netherlands Centre for Indigenous People

The Netherlands Centre for Indigenous Peoples (NCIV) is a non-governmental organization based in Amsterdam which supports the promotion and protection of the rights of indigenous peoples worldwide since 1969. One of the main activities of NCIV is the administration of the grants for the rights of indigenous peoples (GRIP). With GRIP, NCIV seeks to increase financial support to small-scale initiatives coming from organizations of indigenous peoples in order to promote, at the local level, the recognition and implementation of their rights. Project proposals and requests for information are to be directed to the Coordinator of GRIP.

Contacts: *NCIV; P.O.Box 94098; 1090, GB Amsterdam, The Netherlands*
Phone: ++31-20-693 8625, Fax: ++31-20-665 2818
http://www.nciv.net

New England Biolabs Foundation

NEBF is an independent private foundation which supports grassroots organizations working with the environment, social change, the arts, elementary education, and science. The Foundation's funding priorities are targeted to environmental issues with special program interests in: marine conservation; estuary protection; sustainable economic development; sustainable organic agriculture; environmental education for teachers or elementary school students.

Contacts: *New England Biolabs Foundation; 8 Enon St. #2B; Beverly, MA 01915 USA*
Phone: 978-927-2404; Fax: 978-998-6837
http://www.nebf.org

NIPPON FOUNDATION

The Nippon Foundation works throughout the world, focusing on developing countries, to help the disadvantaged overcome fundamental barriers such as disease, hunger and poverty, as they strive to attain a life of self-sufficiency. In Africa, the Foundation focuses on three areas: human resources development, basic human needs, and leprosy elimination. The Nippon Foundation also supports a variety of scholarships, particularly in the maritime sciences. See website for complete grant opportunities.

Contacts: *The Nippon Foundation Akasaka 1-2-2, Minato-ku, Tokyo*
Phone: 81-3-6229-5111
http://www.nippon-foundation.or.jp

NOUVELLE PLANETE

Nouvelle Planete works with local organizations and grassroots groups to design and implement community development projects. In the field of agriculture NP supports initiatives like: ecological agriculture courses; construction of wells, dams, dykes and irrigation canals to develop cultures in the dry season, and to fight desertification; creation of tree nurseries and orchards in villages, and of reforested areas; enhancing the sustainable exploitation of natural resources. With respect to appropriate technology NP supports: training courses and local production of appropriate technology for villages and farmers; initiatives in the field of solar energy and other renewable energy sources to replace traditional fuels like wood.

Contacts: *Director, Chemin de la forêt 3, 1042 Assens – Switzerland,*
Phone: +41-21-881.23.80; Fax: +41-21-882.10.54
http://www.nouvelle-planete.ch

NETHERLANDS ORGANIZATION FOR INTERNATIONAL DEVELOPMENT COOPERATION

NOVIB has over forty years of experience with thousands of projects in the Third World covering a wide range of areas and subjects. It has evolved a way of working called the 'Novib Method'. The Novib Method involves working to realize the structural, sustainable eradication of poverty in Africa, Asia, Latin America and Eastern Europe. Novib supports projects that focus on such areas as health care, rural development, education, and setting up small businesses. Other priority issues include environment, gender, and human rights.

Contacts: *PO Box 30919, 2500 GX Den Haag - the Netherlands*
Phone: +31-70-3421621;
Fax: +31-70-3614461;
http://www.novib.org

OAK FOUNDATION

The Oak Foundation commits its resources to address issues of global social and environmental concern, particularly those that have a major impact on the lives of the disadvantaged. Focus on the environment emphasizes conservation of marine ecosystems and promotion of energy efficient technologies to reduce global warming. Oak's programme priorities and areas of particular emphasis are: Climate change, human rights, homelessness, learning differences, marine conservation, child abuse, women's issues and special interests.

Contacts: *http://www.oakfnd.org*

ORGANIZATION FOR INDUSTRIAL, SPIRITUAL AND CULTURAL ADVANCEMENT

OISCA-International operates on a conviction that sound development is rooted foremost on agriculture and rural-based primary industries. While promoting no specific religion, OISCA-International stresses a spiritual component along with material development. One of its main activities is the Tree Planting Programme: OISCA has been engaged in creating community forests in several Asian countries. By mobilizing volunteers, mainly from Japan, it cooperates with partner agencies and organizations in Asian.

Contacts: *6-12, Izumi 3-chome, Suginami-ku, 168-0063 Tokyo – Japan*
Phone: +81-3-3322-5161 Fax: +81-3-3324-7111
overseas@oisca.org

ORGANIZATION OF PETROLEUM EXPORTING COUNTRIES FUND FOR INTERNATIONAL DEVELOPMENT

Fund for International Development: The Fund aims to promote cooperation between OPEC member countries and other developing countries as an expression of South-South solidarity and to help particularly the poorer, low-income countries in pursuit of their social and economic advancement. Through its grant window, OFID channels much-needed resources into a wide variety of schemes and activities for which loan assistance is not an option. OFID grants include technical assistance for deserving social causes, funding for research and studies, and emergency relief following natural or man-made calamities.

Contacts: *OPEC Fund for International Development*
Parkring 8, A-1010, Vienna, Austria
info@ofid.org; http://www.opecfund.org

OROVERDE

The full name is OroVerde Stiftung zur Rettung der Regenwälder which means Foundation for the Rescue of Rainforests. OroVerde's aim is to foster the protection and conservation of tropical forests in combination with sustainable development practices. It provides funds for initiatives regarding: establishing protected areas, reforestation measures, introduction of forest restoration mechanisms, and environmental education.

Contacts: *Radilostraße 17-19, 60489 Frankfurt - Rödelheim – Germany*
Phone: +49-69-609195-0; Fax: +49-69-620979
info@oroverde.de
http://www.oroverde.de

Overbrook Foundation

The Overbrook Foundation strives to improve the lives of people by supporting projects that protect human and civil rights, advance the self-sufficiency and well being of individuals and their communities and conserve the natural environment. In pursuit of its mission, the Foundation advances programs in the Environment and Human Rights. The Environment Program supports organizations working to develop better consumption and production habits.

Contacts: *The Overbrook Foundation; 122 East 42nd St., Suite 2500; New York, NY 10168 USA*
Phone: (212) 661-8710; Fax: (212) 661-8664
http://www.overbrook.org

Oxfam

Oxfam's work is prioritized according to five strategic change objectives relating to (1) sustainable livelihoods, (2) health and education, (3) life and security, (4) the right to be heard, and (5) gender equity. Oxfam GB's programmes are mostly implemented by local partner organizations in developing countries. All funding related matters are dealt with by Oxfam GB through its international field offices. Besides financing development projects in developing countries, Oxfam GB, together with the Oxfam International family, are involved in campaigning and lobbying activities, such as the Trade Campaign. Many of the projects Oxfam funds explicitly support sustainable natural resources management. Also, potential environmental impacts of a proposed project are considered before funding is approved.

Contacts: *Oxfam House, 274 Banbury Road, OX2 7DZ Oxford - United Kingdom*
Phone: +44-1865-312.610
http://www.oxfam.org.uk

Pacific Environment

Founded in 1987 as the Pacific Environment and Resources Center (PERC) this non-profit organization started by working on international energy and resources issues. Pacific Environment protects the living environment of the Pacific Rim by strengthening democracy, supporting grassroots activism, empowering communities, and redefining international policies.

Contacts: *1440 Broadway, suite 306, CA 94612, Oakland – USA*
Phone: +1-510-2518800; Fax: +1-510-251.8838
info@pacificenvironment.org
http://www.pacificenvironment.org

Patagonia Inc.

Patagonia is an outdoor clothing company that gives grants for environmental work. Patagonia funds only environmental work. They are most interested in making grants to organizations that identify and work on the root causes of problems and that approach issues with a commitment to long-term change. Patagonia looks for programs with a clear agenda for change and a strategic plan for achieving the organization's goals.

Contacts: *PO Box 150, CA 93002, Ventura – USA*
info@patagonia.com
http://www.patagonia.com

Pepsico Foundation

The PepsiCo Foundation invites grant proposals in the following areas: health and wellness; diversity and inclusion; environment: to advance the knowledge and methods of water resource; management which are sustainable and positively impact both quantity and quality of water supply in developing nations; thought leadership.

Contacts: *http://www.pepsico.com*

Phillip Morris International

Contributes to charitable, nonprofit organizations that fight social problems such as hunger and poverty, provide disaster relief, and support victims of domestic violence and foster cultural activities around the world.

Contacts: *Philip Morris International Management SA; 107 Avenue de Court; 1001, Lausanne Switzerland*
Phone: +41 (0)21 618 6111
http://www.philipmorrisinternational.com

Plant-IT 2000

This is a non-profit foundation dedicated to properly planting, maintaining and protecting as many indigenous trees as possible worldwide. In addition to replenishing cities and forests, Plant-It 2000 also periodically provides educational workshops on forestry issues to thousands of children. Founded by the late Singer John Denver in 1992, its reforestation projects in many countries often directly help humanity by increasing crop production, fresh water availability and food.

Contacts: *PMB 310, 9457 S. University Boulevard, Highlands Ranch, CO 80126 – USA*
Phone: +1-303-2210077; Fax: +1-303-2210090
trees@plantit2000.com
http://www.plantit2000.com

POVERTY AND ECONOMIC POLICY RESEARCH NETWORK

The Poverty and Economic Policy (PEP) Research Network brings together and provides support to developing country researchers working to reduce poverty. The network receives funding from the International Development Research Centre (IDRC). The priority goals are to: better understand the causes and consequences of poverty; propose alternative strategies, policies and programs to reduce poverty; improve the monitoring and measurement of poverty; develop local research (and training) capacity in these areas; develop new concepts and methodologies through fundamental research. PEP is composed of three tightly linked constituent networks: Community-Based Monitoring Systems (CBMS) - designs and pilots community-based monitoring and local development systems aimed at poverty in its multidimensional sense.

Contacts: *http://www.pep-net.org*

POVERTY REDUCTION AND ENVIRONMENTAL MANAGEMENT

The Poverty Reduction and Environmental Management (PREM) programme aims to deepen and broaden the exposure of economic researchers and policy advisors in developing countries to the theory and methods of natural resource management and environmental economics. It is envisaged that this will encourage effective policy change in developing countries with the joint goals of poverty reduction and sustainable environmental management. PREM would rather fund projects with some preparatory basis or existing background research, but which have not yet explored the related environmental economic issues. An example might be a road construction project in a national park, where ecologists have conducted an environmental impact assessment and engineers have figured out the most appropriate route.

Contacts: PREM Programme Officer; Institute for Environmental Studies (IVM),
http://www.prem-online.org

PRESBYTARIAN HUNGER PROGRAMME

PHP provides grants to programs addressing hunger and its causes. One of the five program areas is development assistance for socio-economic activities. PHP supports projects that encourage and support land ownership by the poor, appropriate agricultural technology, rural community development, effective soil conservation, water resource development, and community forestry.

Contacts: *100 Witherspoon Street, KY 40202-1396 Louisville – United States*
Phone: +1-502-569.8963
http://www.pcusa.org

PRICEWATERHOUSE COOPERS ENDOWMENT FOR THE BUSINESS OF GOVERNMENT

Through grants for research and forums, the PricewaterhouseCoopers Endowment for the Business of Government stimulates research and facilitates discussion on new approaches to improving the effectiveness of government at the federal, state, local, and international levels. Research grants provide funding for individuals in universities, nonprofit organizations, and journalism to prepare research papers. Individuals receiving grants are responsible for producing a 20-40 page research report on one of the Endowment's approved topics. The reports are published and disseminated by the PricewaterhouseCoopers Endowment.

Contacts: *The PricewaterhouseCoopers Endowment for the Business of Government; 1616 North Fort Myer Drive; Arlington, VA 22209*
http://www. endowment.pwcglobal.com

PUBLIC WELFARE FOUNDATION

This is a non-government grant making organization dedicated to support organizations that provide services to disadvantaged populations. The Foundation focuses its work on addressing the human health impact of environmental degradation and pollution, particularly on disadvantaged communities or those with the least resources to address this impact. Emphasis is given to such issues as air, water, and land contamination. Other themes supported are sustainable development, advocacy, and policy development.

Contacts: *99 Park Avenue, Suite 2220, NY 10016, New York – USA*
Phone: +1-212-3701165; Fax: +1-212-599.6282
http://www.publicwelfare.org

RAINFOREST ACTION NETWORK

RAN works to protect the Earth's rainforests and support the rights of their inhabitants through education, grassroots organizing, and non-violent direct action. RAN works in alliance with environmental and human rights groups around the world, including indigenous forest communities and non-governmental organizations in rainforest countries. From its beginnings, RAN has played a key role in strengthening the worldwide rainforest conservation movement by supporting activists in tropical countries, as well as by organizing and mobilizing consumers and community action groups throughout the United States.

Contacts: *221 Pine St., Suite 500, CA 94104, San Francisco – USA*

Phone: +1-415-398.44.04; Fax: +1-415-398.27.32
rainforest@ran.org
http://www.ran.org

RIGHT SHARING OF WORLD RESOURCES

RSWR is a Quaker organization supported by the Religious Society of Friends (Quakers) in the United States. RSWR has two main goals: *(i)* provide grants to new, small grassroots organizations in the developing world; *(ii)* Help Quakers in the United States learn about the negative effects of poverty in the developing world and the negative effects of materialism in North America. They provide seed grants to organizations to implement an income generating project, offering a means of establishing self sufficiency to very poor people. They strive to help newer, smaller organization whose budgets may not allow them to receive grants from other sources.

Contacts: *Right Sharing of World Resources;*
101 Quaker Hill. Dr.; Richmond, IN 47374 USA,
Phone: 765-966-0314, Fax: 765-966-0325
rswr@rswr.org
http://www.rswr.org

ROLEX AWARDS FOR ENTERPRISE

Rolex Awards recognizes pioneering concepts and innovative thoughts by giving individuals the means to carry out a major undertaking. The Rolex Awards fund new or ongoing work. The Awards support five areas of human endeavor: science and medicine; technology and innovation; exploration and discovery; the environment; and cultural heritage.

Contacts: *Rolex Awards for Enterprise; P.O. Box 1311; 1211, Geneva 26 Switzerland*
Phone: +41 22 302 22 00, Fax: +41 22 302 25 85
secretariat@rolexawards.com
http://www.rolexawards.com

SAHEL DEFIS

Developpement Environnement Formation Insertion Sociale - short Sahel Défis - represents a network of Sahelian NGOs active in the field of desertification control and poverty alleviation. The goal is to provide financial and technical support to local initiatives, as a first step towards major sustainable development.

It takes advantage of the CCD process to launch, support and follow up on field activities initiated by local NGOs in various countries of the Sahel. Sahel Dàfis supports local initiatives in environmental protection and income generation for women and youth in the Sahel region of Africa.

Projects focus on training, income and employment generation, natural resource management and the rehabilitation of degraded natural sites.

Contacts: *253 chemin de Fontanières, 69350 La Mulatière – France*

Phone: +33-4- 7851.4888; Fax: +33-4-7851.4858

http://www.saheldef s.free.fr

SAMUEL RUBIN FOUNDATION

The Foundation's general purpose is to carry on the vision of its founder, Samuel Rubin, whose life was dedicated to the pursuit of peace and justice and the search for an equitable reallocation of the world's resources. The Foundation believes that these objectives can be achieved only through the fullest implementation of social, economic, political, civil and cultural rights for all the worlds' people.

Contacts: *Grants Administrator; Samuel Rubin Foundation; 777 United Nations Plaza; New York, NY 10017-3521 USA*

Phone: 212 697-8945, Fax: 212 682-0886

http://www.samuelrubinfoundation.org

SASAKAWA PEACE FOUNDATION

The Sasakawa Peace Foundation sponsors a wide range of programs designed to address international issues and to promote global peace via nonprofit sector initiatives worldwide. Founded in September 1986, SPF is a non-profit organization aiming to contribute to the welfare of humankind and development of a sound international community. The Sasakawa Pacific Island Nations Fund, SPINF, accepts proposals and supports projects in areas of people exchange and human resource development that adhere to the fund's two objectives: *(i)* the promoting of interpersonal exchange activities between the peoples of Japan and of the Pacific island nations; and *(ii)* the building of linkages between government bodies, NGOs and other organizations interested in strengthening harmony within the region and in enhancing the quality of people's lives in its island societies.

Contacts: *The Nippon Foundation Bldg 1-2-2 Akasaka, Minato-ku, Tokyo 107-8523, Japan*

Phone: +81 (3) 6229-5400; Fax: +81 (3) 3769-2090

http://www.spf.org

SEED INITIATIVE

The Seed Initiative (Supporting Entrepreneurs in Environment and Development) aims to inspire, support and build the capacity of locally driven entrepreneurial partnerships to contribute to the delivery of the Millennium Development Goals. The initiative focuses on 'business as unusual' – innovative action delivering real solutions through project cooperation among small and large businesses, local and international NGOs, women's groups, labour organizations, public authorities and UN agencies, and others working in the field of sustainable development. The Seed Awards, an international biennial competition, is designed to support innovative, entrepreneurial partnerships. Through the provision of targeted services, support, training and exposure to additional networks, the Seed Initiative helps young and promising partnerships bring their initiatives to fruition. The award is not monetary but a comprehensive package of tailor-made services from a range of organizations.

Contacts: *Manager, Seed Secretariat, c/o IUCN; rue Mauverney 28; CH - 1196 Gland Switzerland*
Phone: +44 79 77 218057
http://www.seedinit.org

SEEDTREE

This small non-profit organization was founded in 1995, it assists self-reliant efforts to protect and renew forests and ecosystems through human economy and ecology, particularly by tree planting, water supply and biogas. SEEDTREE's aim is to preserve and renew forest ecosystems by developing a sustainable human ecology.

Contacts: *RR 2 Box 802 West Cape, ME 04981 Stockton Springs – USA*
Phone: +1- 207-567-3056
http://www.seedtree.org

SELF HELP DEVELOPMENT INTERNATIONAL

Self Help works by helping people to help themselves to recover their independence and self-sufficiency. Food aid may be essential in the short term, but in the long term the only way to break the famine cycle is through development. SHDI implements rural development programmes with components like conservation and restoration of the natural resources including measures against soil erosion and land degradation (afforestation, conservation), crop production (seed distribution, irrigation), water distribution, livestock development, women, health, education/training, capacity building.

Contacts: *Hacketstown, Co. Carlow – Ireland*
Phone: +353-508-71175; Fax: +353-508-71292
http://www.selfhelpintl.ie

SIEMENPUU FOUNDATION

The Siemenpuu Foundation supports people in the South to get their voices heard and at the same time to support their work in advancing citizens' political and other decision-making powers locally and globally; in protecting biological and cultural diversity; in securing the safety of their environment and control over natural resources, and in furthering ecologically sustainable production and consumption. Primarily, the Foundation gives support directly to civil society organizations and networks of organizations, NGOs, community groups and research institutes operating and based in developing countries. The nature of the activities supported can be diverse: projects, research, publications, training, environmental and human rights advocacy work (both locally and globally), and support for communities in getting organized and politically empowered etc.

Contacts: *Aleksanterinkatu 48 A, 00100 Helsinki – Finland*
Phone: +58-9-272.23.36; Fax: +358-9-622.718.92
info@siemenpuu.org
http://www.siemenpuu.org

SOS Sahel

SOS Sahel is the only British development organization that focuses exclusively on the semi-arid lands of the Sahel and Horn of Africa. Projects are designed by local people to meet their needs and address their problems.

Contacts: *10c Littlegate Street, St. Ebbes, OX1 1QT Oxford - United Kingdom*
Phone: +44-1865-723.200; Fax: +44-1865-722.355
http://www.sahel.org.uk

Sustainable Harvest International

The mission of SHI is to reverse environmental degradation by helping rural inhabitants in Central America restore ecological stability and sustainable economic productivity to overexploited lands. SHI facilitates farmers and communities in the tropics with long-term assistance implementing environmentally and economically sustainable technologies and land use practices that alleviate poverty by restoring ecological stability.

Contacts: *81 Newbury Neck Rd., ME 04684 Surry - United States*
Phone: +1-207-664.0987; Fax: +1-207-664.0700
info@sustainableharvest.org
http://www.sustainableharvest.org

Swedish International Development Agency

SIDA Scholarships cover: Sida's travel grant for internships; minor field studies (MFS); and Linnaeus-Palme (an exchange programme financed by Sida for university teachers and students. The first part, Linnaeus, enables Swedish teachers and students to study in the third world. The second part, Palme, offers teachers and students in developing countries the chance to come to Sweden. Support to NGOs includes: Funds are allocated to the aid operations of charities and non-governmental organizations in developing countries.

Contacts: *Valhallavägen 199, 10525 Stockholm Sweden*
Phone:+ 46 8 698 50 00, Fax:+46 8 20 88 64
sida@sida.se
http://www.sida.org

Swiss Agency for Development and Co-operation

SDC is part of the Swiss department of foreign affairs that executes the government's development and humanitarian aid. SDC concentrates its long-term development co-operation efforts on a limited number of countries and areas of activity. SDC concentrates its development efforts on poverty alleviation, sustainable use of natural resources, supporting the empowerment of people, gender equality, and providing access to information. They also support initiatives to protect endangered areas. Current priorities of the SDC include safeguarding and promoting peace and security, human rights, democracy and rule of law. Swiss development co-operation aims above all at improving the productive base, by fostering agriculture and food security, together with industrial and crafts production and the development of appropriate technologies. In particular, it will give financial support to socially feasibly structural adjustment programs.

Contacts: *Freiburgstrasse 130, 3003 Berne – Switzerland*
Phone: +41-31-322.34.75 ; Fax: +41-31-324.13.48
info@deza.admin.ch
http://www.sdc.admin.ch

Swiss Aid

Swiss Aid supports development initiatives of local partner organizations. It seeks to support projects that promote the autonomy of local populations' and it encourages people to develop sustainable solutions to their problems on the basis of their own knowledge, creativity, resources and traditions. SwissAid supports all aspects of rural development including, agriculture, erosion protection, irrigation and sanitation, transport, crafts, education, healthcare, and organizational development. Special attention is given towards the role of women within the frame of regular projects and by supporting special women's projects, the latter which are initiated, conducted and exclusively aimed at women. Other fields of priority are biotechnology and MFI's (national and global). Grants range from medium to large.

Contacts: *Jubiläumsstrasse 60, 3000 Berne – Switzerland*
Phone: +41-31-350.53.53; Fax: +41-31-350.27.83
http://www.swissaid.ch

Threshold Foundation

Threshold Foundation is a membership organization comprised of individuals with significant financial resources who contribute their time, money and energies to support progressive social change through volunteer-based philanthropy. Grant funds are derived from a combination of annual member contributions and payouts from their endowment. The Foundation serves the social change movement by funding innovative national and international non-profit organizations working to further the causes of social justice, environmental preservation, humane economic systems and the peaceful coexistence of individuals, communities and cultures.

Contacts: *Threshold Foundation; PO Box 29903; San Francisco, CA 94129-0903 USA*
Phone: 415-561-6400, Fax: 415-561-6401
http://www.thresholdtides.org

TOYOTA FOUNDATION

The Toyota Foundation is a private, nonprofit, grant-making organization dedicated to the goals of realizing greater human fulfillment and contributing to the development of a human-oriented society. The Foundation administers a research grant program, an initiative grant program, a network formation program, and a grant program for community activities. Themes change yearly; please refer to the webpage for further guidance.

Contacts: T*he Toyota Foundation; 37F, Shinjuku-Mitsui Building; 2-1-1, Nishi-shinjuku, Shinjukuku, Tokyo, Japan*
Phone: 81-3-3344-1701, Fax: 81-3-3342-6911
http://www.toyotafound.or.jp/etop.htm

TRACE FOUNDATION

The mission of the Trace Foundation is to support economically sustainable and environmentally sound initiatives that strengthen communities and develop knowledge and skills. The goals of the foundation include improving education and training, increasing employment opportunities for the poor, conservation and management of natural resources (including recycling and refuse management, forestry and grassland management and solar and wind energy), and expansion of health care in rural areas. The foundation views strengthening capacity through training as a primary goal of all its programs.

Contacts: *31 Perry Street, NY 10014 New York – USA*
Phone: +1-212-367.7380; Fax: +1 212-367.7383
http://www.trace.org

TREE AID

TREE AID is working to alleviate poverty and reverse environmental degradation through community forest projects in arid Africa. It aims to strengthen the long-term self-reliance of people in Africa's drylands, so that they can meet their basic needs, earn an income, and increase the stability of their communities through the regeneration of their woodlands. TREE AID does not implement its own projects. Rather it funds local organizations that are already working with village communities they know. However, TREE AID does provide training to strengthen the technical and management capacity of the partner organizations, so that they can run their own projects long after funding has ended.

Contacts: *Brunswick Court, Brunswick Square, BS2 8PE Bristol - United Kingdom*

Phone: +44- 117-909.6363; Fax: +44-117- 909.6617
http://www.treeaid.org.uk

TREES FOR LIFE

Trees for Life is a people-to-people movement that helps plant fruit trees in Cambodia, India and Latin America. Each fruit tree protects the environment and provides a low-cost, self-renewing source of food for a large number of people. Trees for Life's activities include three elements: education, health and environment. Volunteers in the villages are trained by Trees for Life and provided with essential support for programs that respond to the needs of their communities. Besides fruit trees the organization also supports: introduction of fuel-efficient cook-stoves, construction of water wells, and measures for soil regeneration in order to improve degraded land.

Contacts: *3006 W St Louis, KS 67203 Wichita-United States*
Phone: +1-316-945.6929; Fax: +1-316-945.0909
http://www.treesforlife.org

TREES FOR THE FUTURE

This is a non-profit organization with a people-to-people, action program that promotes and supports tree planting and environmentally sustainable self-help projects, in co-operation with local people in their own (sub-tropical) communities around the world. Their mission is to reverse forest decline (moist, cloud, rain, dry, etc.), to improve the quality of life for those who are dependent upon the land and forests, and to ensure healthy forests for future generations, by assisting local people to employ sustainable land management practices through the means of in-kind donations of tree seeds, hand tools and planting materials, as well as on-site technical training and planning assistance. Other issues they give emphasis to include: reforestation, afforestation, biological diversity, desertification and the restoration of degraded lands.

Contacts: *PO Box 7027, MD 20907, Silver Spring – USA*
Phone: +1-301-565.0630; Fax: +1-301-565.5012
http://www.treesftf.org

TROPICAL RAINFOREST COALITION

TRC is an all-volunteer, non-profit organization that concentrates on the preservation of rainforests and indigenous peoples. It builds coalitions between individuals, corporations, businesses, learning institutions and NGOs that are working to reverse the unsustainable use of rainforests. Funds are small and priority is given to the preservation and conservation of tropical rainforest ecosystems and the indigenous peoples who live in those systems. Other priorities include funding and managing land purchase efforts (Save-an-Acre Program), developing low-impact eco-tourism programs with indigenous participation, promoting environmental awareness and education, and the Tropical Bird Rescue program.

Contacts: *21730 Stevens Creek Blvd., Suite 102, CA 95014, Cupertino – USA*
http://www.rainforest.org

UC IGCC

Understanding international conflict and promoting cooperation among nations in political, economic, and environmental affairs causes of international conflict; the resolution of international disputes; international and regional cooperation on security, economic, health, environmental, cultural, legal, and social issues; the development and operations of regional and international organizations; the economics, politics and sociology of transnational flows of capital, goods, technology, ideas, and people; transnational social movements and non-governmental organizations; and gender issues and international politics.

Contacts: *IGCC; University of California, San Diego 9500 Gilman Drive, Dept.0518 LaJolla, CA 92093-0518*

UNDP: The Equator Initiative

One prize is awarded in each region of eligibility (Latin America and the Caribbean, Africa, and Asia and the Pacific), one to a community-based project in a UNESCO World Heritage Site, and one to the project that best exemplifies sustainable biodiversity-based business. The Equator Initiative is a partnership that brings together the United Nations, civil society, business, governments and communities to help build the capacity and raise the profile of grassroots efforts to reduce poverty through the conservation and sustainable use of biodiversity. The Equator Prize is a prestigious international award that recognizes outstanding local efforts to reduce poverty through the conservation and sustainable use of biodiversity.

Contacts: *Equator Initiative; 405 Lexington Avenue, 5th Floor; New York, NY 10174 USA*
Phone: 212-457-1709, Fax: 212-457-1370
http://www.EquatorInitiative.org

USAID

The United States Agency of International Development is the government's agency responsible for implementing the United States' international development and foreign humanitarian assistance programs. There are several mechanisms currently available for USAID to provide support to forestry and other environmental activities, including an Interagency Agreement between USAID and the US Forest Service (International Programs), several "Indefinite Quantities Contracts" the most recent being the "BIOFORIQC" which is with two major US international development consulting firms and their partners. The main partner for the USAID Global Bureau Forestry Team is the US Forest Service, International Programs.

Contacts: *Ronald Reagan Building, DC 20523-1000, Washington DC –USA*
Phone: +1-202-712-4810; Fax: +1-202-216-3524
http://www.usaid.gov

VISKOGEN

The Foundation ViSkogen started in 1983 with a cry in the magazine 'Vi' to help stop the spread of the desert in Africa. The focus of the Vi Agro-forestry Programme is the preservation of trees in East Africa. ViSkogen plants millions of trees every year, to prevent famine and help East Africans build themselves a sustainable future.

Contacts: *PO Box 302 27, 10425 Stockholm – Sweden*
Phone: +46-8-657.47.36; Fax: +46-8-657.47.39
plantera@viskogen.se
http://www.vi-skogen.com

VODAPHONE FOUNDATION

The Foundation makes social investments that help the people of the world to have fuller lives by: sharing the benefits of developments in mobile communications technology as widely as possible; protecting the natural environment; and supporting the local communities in which Vodafone's customers, employees, investors and suppliers live. Most projects are funded in partnership with local Vodafone company foundations or community programmes.

Contacts: *http://www.vodafonefoundation.org*

VOLKSWAGEN FOUNDATION

The Volkswagen Foundation focuses its efforts on selected funding initiatives. In special circumstances the Foundation is also prepared to support extraordinary projects which fall outside the scope of its defined funding activities.

Contacts: *The Humanities and Social Sciences*
Phone: +49 (0)511/83 81-276; Fax: +49 (0)511/83 81-344;
http://www.volkswagenstiftung.de

VOLVO ENVIRONMENT PRIZE

The Volvo Environment Prize was formally instituted in May 1988 by the Volvo Annual Shareholders Meeting, with the objective of promoting research and development across the environmental spectrum, by acknowledging people who have made an outstanding contribution to understanding or protecting the environment through scientific, socioeconomic or technological innovation or discovery of global or regional importance. The Volvo Environment Prize is awarded for "Outstanding innovations or discoveries scientific, socio-economic, or technological which have direct or indirect significance in the environmental field and are of global or regional importance"

Contacts: *Volvo Environment Prize Foundation; Volvo Group, Headquarters;1650 VHK; S-405 08 Göteborg Sweden*
Phone: +46 31 66 91 04; Fax: +46 31 53 8431
http://www.environmentprize.com

W.K. Kellogg Foundation Africa Programme

International Fellowships (Southern Africa). Botswana, Lesotho, Mozambique, South Africa, Swaziland, and Zimbabwe. An integrated approach is taken to addressing the Foundation's four programming interests. Following are the strategies that merge these interests: organizational and institutional development and transformation; strengthen leadership capacity; strengthen the capacity of communities. Candidates must be nominated by their organization for these long or short-term study grants. The Kellogg Foundation does not generally consider requests from people already in the U.S

Contacts: *W.K. Kellogg Foundation Africa Program Office, Private Bag X 36 Hatfield Pretoria, 0028, Republic of South Africa*
Phone: +27-12-431-0900; Fax: +27-12-342-3617
http://www.wkkfafrica.org

Wallace Global Fund

The Fund supports activities at the global and national level, and considers significant local or regional initiatives offering the potential to leverage broader national or global impact. The objectives of the fund include: effective protection of the environment and natural resources and their capacity to provide for human needs; progressive public policies that serve the common good, protect civil liberties, and guard against corporate abuses; sustainable levels of human populations; and equal justice.

Contacts: *Wallace Global Fund; 1990M Street, NW, Suite 250; Washington, DC 20036 USA*
Phone: (202)452-1530, Fax: (202)452-0922
http://www.wgf.org

Weeden Foundation

The Foundation has prioritized the protection of biodiversity. Organizations supported to date range from those that protect ecosystems and wildlife to those that raise the status of women and increase awareness about family planning. The Foundation tends to provide large funds to projects, which serve as catalysts and focus on native forest conservation aimed at promoting greater efficiency in the use of wood products and developing substitutes to wood fibre. Additionally, the Foundation is interested in supporting effective consumer-based and consumer-targeted education and empowerment programs, especially those that use unique and effective delivery systems and outreach methods. The Weeden Foundation considers funding in any geographical location, but the Trustees have chosen the western US, Chile, Bolivia, and Central Siberia as high priority. Applicants should review the descriptions of past grantees posted on their website to better clarify the types of programs and organizations the Foundation typically funds. Funding outside of priority areas is rare.

Contacts: *747 Third Avenue, 34th Floor, NY 10017, New York – USA*
Phone: +1-212-888-1672; Fax: +1-212-888-1354
http://www.weedenfdn.org

WELTFRIEDENSDIENST

The Weltfriedensdienst, what literally means "world peace service", was established in 1959. For the WFD, peace is more than just the absence of war. In the projects the WFD supports, securing the basis of existence is a main concern. In the international field WDT concentrates on environmental protection, the conservation of natural resources and sustainable development in combination with the support of grass-roots groups and local initiatives in Southern and West Africa.

Contacts: *Hedemannstr.14, 10969 Berlin – Germany*
Phone: +49-30-253.990.0; Fax: +49-30-251.18.87
http://www.wfd.de

WHITLEY FUND FOR NATURE

Whitley Awards were developed to recognize conservation leaders from around the world who are applying sustained effort to conserve the natural environment. The scope of interest has in the past included, but is not limited to: protecting endangered ecosystems and species, promoting sustainability, and influencing environmental policies. Successful applicants are often supported by a strong team and/or local NGO which exhibits a broad range of relevant skills and knowledge. The inclusion of locals as team members is considered favourably, however, it is also understood that in some situations, the inclusion of international team members with specialist skills and knowledge may be required for project success. Community and stakeholder education is considered very important to successful nature conservation projects. WFN funds initiatives that aim to address broad nature conservation and sustainability issues, and which seek long-term solutions to environmental problems. Projects that take an ecosystem approach are therefore favored over species specific work.

Contacts: *The Whitley Fund for Nature; 50 Queensdale Rd.; London W11 4SA, United Kingdom*
Phone: +44 (0)20 7602 3443
http://www.whitleyaward.org

WINDSTAR FOUNDATION

The Windstar Foundation, co-founded by John Denver and Thomas Crum, has consistently sought ways, "To inspire individuals to make responsible choices and take direct personal actions to achieve a peaceful and environmentally sustainable future." With this mission in mind, the foundation is awarding two scholarships for qualified undergraduates entering their junior or senior year of college, as well as one scholarship for graduate students entering their second year of graduate school. The applicant must be a U.S. student enrolled in an accredited U.S. college or university Environmental Studies or Environmental Engineering Program. Applicants must have a college GPA of 3.0 or better in his or her major field.

Contacts: *http://www.wstar.org/Windstar/Programs/*

WINGS GLOBAL FUND FOR COMMUNITY FOUNDATIONS

Community foundations are important vehicles for community-driven development and capacity building. The Global Fund for Community Foundations is a new WINGS program created to promote the growth and sustainability of community foundations in developing and transitioning countries. The program provides grants primarily to community foundations, though in some instances support organizations will also be eligible.

Contacts: *WINGS Global Fund for Community Foundations*
C/o European Foundation Centre; 51, Rue de la Concorde; 1050 Brussels Belgium
Fax to: + 32 (2) 508 3789
http://www.wings-globalfund.org

WINROCK INTERNATIONAL

The Winrock International Livestock Research and Training Center was created to improve animal agriculture worldwide. The merger with the Agricultural Development Council (founded by John D. Rockefeller) and the International Agricultural Development Service led to the creation of the Winrock International Institute for Agricultural Development. The mission of Winrock is to work with people all around the world in order to increase economic opportunity, sustain natural resources, and protect the environment. Winrock matches innovative approaches in agriculture, natural resources management, clean energy, and leadership development with the unique needs of its partners.

Contacts: *2101 Riverfront Drive, AR 72202 Little Rock - United States*
Phone: +1-501-280.3000
http://www.winrock.org

WOLF FOUNDATION: THE WOLF PRIZES

Since 1978, five or six prizes have been awarded annually in the Sciences. Prize fields comprise: agriculture, chemistry, mathematics, medicine and physics. In the Arts, the prize rotates annually among architecture, music, painting and sculpture. In the event of two or three recipients sharing the prize, the honorarium is divided equally. International prize committees in each field, comprising renowned experts, select the Wolf Prize winners. Prize committees are appointed every year.

Contacts: *The Wolf Foundation; P.O. Box 398; Herzlia Bet 46103, Israel*
Phone: 972-9-955 7120; Fax: 972-9-954 1253
http://www.wolffund.org.il

WOMEN'S WORLD SUMMIT FOUNDATION: PRIZE FOR WOMEN'S CREATIVITY IN RURAL LIFE

The Prize aims to draw international attention to laureates' contributions to sustainable development, household food security and peace, thus generating recognition and support for their projects. While rural women are vital in providing examples of sound practice in their communities, they still do not have full access to tools needed for development, such as education, credit, land rights and participation in decision making. By highlighting and awarding creative development models, innovations and experiences enhancing the quality of rural life, WWSF participates in addressing the eradication of rural poverty, gender mainstreaming and women's empowerment

Contacts: *WWSF Women's World Summit Foundation; 11 avenue de la Paix;*
1202, Geneva, Switzerland
http://www.woman.ch

WORLD BANK DEVELOPMENT MARKET PLACE GRANT

Development Marketplace (DM) is a competitive grant program of the World Bank that funds innovative, small-scale development projects. The DM Grant Competitions are organized at two levels: global and country/regional competitions. The Global Development Marketplace is held every 12-18 months. Competition themes address pressing development challenges and change with each competition. The call for proposals is open to participants worldwide.

Contacts: *http://www.developmentmarketplace.org*

WORLD BANK POST CONFLICT FUND

The Post-Conflict Fund (PCF) of the World Bank was established in 1997 to enhance the World Bank's ability to support countries in transition from conflict to sustainable peace and economic growth. The Post-Conflict Fund makes grants to a wide range of partners (institutions, nongovernmental organizations, United Nations agencies, transitional authorities, governments, and other civil society institutions) to provide earlier and broader World Bank assistance to conflict-affected countries. Grants are focused on the restoration of the lives and livelihood of war affected population, with a premium placed on innovative approaches to conflict, partnerships with donors and executing agencies and leveraging resources through a variety of funding arrangements.

Contacts: *http://www.worldbank.org*

World Bank Small Grants Program

The purpose of the Small Grants Program is to strengthen the voice and influence of poor and marginalized groups in the development processes, thereby making these processes more inclusive and equitable. Thus, it supports activities of civil society organizations whose primary objective is civic engagement of the poor and marginalized populations. By involving citizens who are often excluded from the public arena, and by increasing their capacity to influence policy and program decisions.

Contacts: *http://www.worldbank.org*

World Bank: Japan Social Development Fund

The Japan Social Development Fund (JSDF) was established by the Government of Japan and the World Bank in June 2000 as an untied mechanism for providing direct assistance to the poorest and most vulnerable groups in eligible member countries of the World Bank Group. The main purposes of the JSDF are to: support innovative programs which directly respond to the needs of the poorest and most vulnerable groups in society; provide rapid and demonstrable benefits, which can be sustained, to the poorest and most vulnerable groups; and build capacity, participation and empowerment of civil society, local communities and NGOs.

Contacts: *Head, Trust Fund Program Administration; Trust Fund Operations*
http://www.go.worldbank.org

World Challenge

World Challenge in association with Shell is a competition aimed at finding individuals or groups from around the world who have shown enterprise and innovation at a grass root level. The hallmark of World Challenge is viewer involvement, casting a global net for ideas from individuals who know of an example of best practice they want to be recognized. The twelve chosen projects will be broadcast within the World Challenge series and viewers/readers can vote for their favorite.

Contacts: *http://www.theworldchallenge.co.uk/index.php*

WORLD FOOD PRIZE

The World Food Prize is the foremost international award recognizing –without regard to race, religion, nationality, or political beliefs – the achievements of individuals who have advanced human development by improving the quality, quantity or availability of food in the world. The Prize recognizes contributions in any field involved in the world food supply -- food and agriculture science and technology, manufacturing, marketing, nutrition, economics, poverty alleviation, political leadership and the social sciences.

Contacts: *The World Food Prize Foundation;*
666 Grand Ave., Suite 1700; Des Moines, IA 50309 USA
Phone: 515-245-3783; Fax: 515-245-3785
http://www.worldfoodprize.org

WORLD LAND TRUST

The World Land Trust seeks to conserve the World's tropical forests, as well as other biologically important habitats, which are threatened by destruction. Its aims are: (i) To protect and sustainably manage natural ecosystems of the world, to conserve their biological diversity, with emphasis on threatened habitats and endangered species; (ii) to develop partnerships with local organizations and organizations to engage support and commitment among the people who live in project areas; and raise awareness in the UK and elsewhere, of the need for conservation, to improve understanding and generate support through education, information and fundraising.

Contacts: *P.O. Box 27, IP19 8AL Halesworth United Kingdom*
Phone: +44-1986-874422; Fax: +44-1986-874425
http://www.worldlandtrust.org

WORLD NEIGHBORS

This is a non-profit, non-sectarian organization that works to reduce poverty, hunger and sickness by combining technical assistance in agriculture, environmental conservation, health and small business with training aimed at strengthening the ability of communities to work together to solve problems. World Neighbors in collaboration with local governments and local communities has developed community-based approaches to natural resource management that ensures the well-being of the people while conserving the natural environment.

Contacts: *4127 NW 122 Street, OK 73120, Oklahoma City – USA*
Phone: +1-800-242.63.87; Fax: +1-405-752.93.93
http://www.wn.org

World Vision

World Vision is a Christian relief and development organization dedicated to helping children and their communities worldwide reach their full potential by tackling the causes of poverty. World Vision has several focus areas, two of which are: *(i)* agriculture and food production: The goal is to increase food security in food deficit regions through agriculture, reforestation, and soil and water conservation programs. Ensuring a safe water supply forms a foundation for almost every project. Besides drilling wells and installing pumps, World Vision trains communities in system maintenance, hygiene, sanitation and in the use of water in vegetable production, aquaculture and agriculture; *(ii)* natural resource management: World Vision considers an on-going commitment to natural resource management as a foundation of all development projects. Efforts include soil and water conservation, livestock management, reforestation, and training in environmentally sensitive land and water management.

Contacts: *World Vision International Programs, 300 "I" Street NE, DC 20002 Washington –US*

Phone: +1-202-572.-6300

http://www.worldvision.org

Zayed International Prize for the Environment

Zayed International Prize for the Environment was established in honour of a unique statesman and a dedicated environmentalist, His Highness Sheikh Zayed Bin Sultan Al Nahyan, Former President of the United Arab Emirates and Governor of Abu Dhabi. Sheikh Zayed is the man whose vision and unwavering commitment to the environmental cause helped turn the Arabian Desert green. The prize recognizes and promotes major pioneering contributions in the field of environment and sustainable development, in accordance with the development philosophy and vision of the late Sheikh Zayed Bin Sultan Al Nahyan, and in support of global initiatives. The Zayed Prize is awarded in any field of environment activity: global leadership; scientific and technological achievements; environmental action with positive impact on society.

Contacts: *API Tower, Sheikh Zayed Road; P.O. Box 28399 Dubai -United Arab Emirates, Phone: + 971 4 3326666, Fax: +971 4 3326777*

zayedprz@emirates.net.ae

http://www.zayedprize.org.ae

EDUCATION, TRAINING, FELLOWSHIPS AND SCHOLARSHIPS

AAAS ENVIRONMENTAL FELLOWSHIP

AAAS has nearly 120,000 individual and institutional members and 262 affiliates, serving 10 million scientists in fields ranging from plant biology to dentistry. Under the fellowship programme, fellows spend one year, beginning in September, at U.S. Environmental Protection Agency (EPA) in Washington, DC, working on an array of projects relating to science, policy and the environment. The program is coordinated by EPA's National Center for Environmental Research within the Office of Research and Development. Prospective fellows must have a PhD or an equivalent doctoral-level degree by the time of application and must have an excellent scientific or technical background

Contacts: *1200 New York Avenue NW, Washington, DC 20005*

Phone: 202-326-6400

webmaster@aaas.org

http://www.fellowships.aaas.org

AAAS MASS MEDIA SCIENCE

Juniors and seniors enrolled in life, physical, or social sciences. Engineering students are also eligible. AAAS selects 20-30 Mass Media Fellows each summer who work for 10 weeks as reporters, researchers, or production in mass media organizations nationwide. Students who pursue non-technical fields are ineligible. Applicants must provide three letters of recommendation and a three-five page writing sample.

Contacts: *1200 New York Avenue NW, Washington, DC 20005*

Phone: 202-326-6400

webmaster@aaas.org

http://www.fellowships.aaas.org

http://www.nt1-ehrweb.aaas.org

AAUW – FELLOWSHIPS

Social Science Research Council Offers Abe Fellowship Program. The Abe Fellowship is designed to encourage international multidisciplinary research on topics of pressing global concern. The program seeks to foster the development of a new generation of researchers who are interested in policy-relevant topics of long-range importance and who are willing to become key members of a bilateral and global research network built around such topics. It strives especially to promote a new level of intellectual cooperation between the Japanese and American academic and professional communities committed to and trained for advancing global understanding and problem solving. Applicants are invited to submit proposals for research in the social sciences or the humanities relevant to any one or combination of the three themes (Global Issues; Problems Common to Industrial or Industrializing Societies; Issues that Pertain to U.S-Japan Relations). Research proposals bearing on any one of these three themes may address issues related to technology and society; consumption, labor and markets; human security; civil society; bioethics; aging societies and other life-span issues; transnational economic relations; sustainable development and global environmental issues; and foreign assistance. The program provides Abe Fellows with a minimum of 3 and maximum of 12 months of full-time support over a 24 month period. Fellowships are in support of an individual, regardless of whether they are working in a group. Request or download application materials, submit to the NY office.

Contacts: *Social Science Research Council 810 Seventh Avenue, 31st Floor, New York, NY 10019*
Phone: (212) 377-2700; Fax: (212)377-2727;
ranis@ssrc.org
http://www.ssrc.org/abefell.htm
http://www.ssrc.org/fcom1.htm

ACADEMIC STAFF SCHOLARSHIPS - ASSOCIATION OF COMMONWEALTH UNIVERSITIES

CSFP Academic Staff Scholarships are intended to help universities in the developing countries of the Commonwealth enhance the experience and build up the numbers of their locally born academic staff. Awards are tenable at approved institutions of higher learning for a maximum of three years. CSFP Academic Staff Scholarships are normally available for study towards a Masters or Doctoral degree, but may form part of a higher degree programme in the scholar's home. Specific conditions of eligibility are: CSFP Academic Staff Scholarships are open to Commonwealth citizens, permanently resident in a Commonwealth country other than the UK, holding or returning to a teaching appointment in a university in the developing Commonwealth. Candidates should, at the time of nomination, have graduated from a first or Master's degree within the last 10 years be aged 42 or under at time of nomination. Application is by nomination by the executive head of the university on whose permanent staff the nominee serves.

Contacts: *Woburn House, 20-24 Tavistock Square London, WC1H 9HFUnited Kingdom*
Phone +44 207 380 6700; http://www.acu.ac.uk

ADVANCED TRAINING FOR LEADERSHIP AND SKILLS AWARDS

These USAID-funded fellowships provide opportunities for qualified candidates in Africa to pursue academic undergraduate and graduate programs in the United States. Applicants must reside in the home country. The home country MUST have a USAID Mission Office. Awards are not offered in all countries. Please contact the USAID office in your home country.

Contacts: *ATLAS, Chief of Party 380 Lexington Avenue, New York, NY 10168, USA*
Phone: +1(212) 949-5666; Fax: +1(212) 818-9505
atlas@aaionline.org
http://www.atlas.org

AFRICAN CAPACITY BUILDING PROGRAM

The African Capacity Building Foundation (ACBF) was established in February 1991. It was the outcome of collaboration between African governments and the international donor community. The major sponsoring agencies of the Foundation are the African Development Bank (AfDB), the United Nations Development Programme (UNDP) and the World Bank. Project/program proposals both at national and regional are expected to fit within one or more of the Foundation's six core competency areas in capacity building, as follows: economic policy analysis and development management, financial management and accountability, enhancement and monitoring of national Statistics, public administration and management, strengthening of policy analysis capacity of national parliaments, professionalization of the voices of the private sector and civil society. Projects must also have relevance to core capacity needs and enhanced stakeholders' ownership, inclusivity of potential stakeholders/commitment to a participatory process in capacity building and national/regional development, etc. All countries in sub-Saharan Africa are eligible for support by ACBF.

Contacts: *The Executive Secretary; The African Capacity Building Foundation;*
7th and, 15th Floors, Intermarket Life Towers, Cnr. Jason Moyo/Sam NujomaStreet,
P. O. Box 1562, Harare, Zimbabwe
Phone: +(263 - 4) 790398/9, 700208/210, Fax: +(263 - 4) 702915, 738520
root@acbf-pact.org
http://www.acbf-pact.org/index.asp

AFRICAN GRADUATE FELLOWSHIP/AMERICAN UNIVERSITY IN CAIRO

Fellowship (covering tuition costs only) for two academic years. African nationals who have a BA or B.Sc. degree with an excellent academic record of not less than "Very Good", an overall GPA of 3.0 on a 4.0 scale, or the equivalent are eligible to apply. Students must show proficiency in English by either submitting a TOEFL with TWE score of 550 and 4.5 or above, or taking AUC's ELPET exam. Some additional program specific and English language requirements also apply.

Contacts: *The Office of International Graduate Programs; 113 Kasr El Aini Street, P.O. Box 2511; Cairo 11511; Egypt. Tel: 20-2-357-5530; fax: 20-2-355-7565 agf@uns2.aucegypt.edu http://www.aucegypt.edu*

African Regional Post-graduate Programme in Insect Science

This is an international course established to train insect scientists (Entomologists) in East and North East Africa. ARPPIS is a collaborative network project involving ICIPE at Nairobi, Kenya, a number of African Universities and donor agencies for the training of insect scientists to solve Africa's numerous insect pest problems. Some Universities in Africa host sub-regional Centres of this Programme, as centres of excellence, for the training of young entomologists and Addis Ababa University has been selected by the ARPPIS Academic Board as the Sub-regional Centre of East and North East Africa. Applicants must hold a first class or second upper degree in Biology, Agriculture, zoology or a related field and must have taken a basic course in entomology in their undergraduate Programme. A few scholarships provided by the German Academic Exchange Service (DAAD) to the ARPPIS. Possible scholarship from German Academic Exchange Service (DAAD). Applicants must have a B.Sc in biological sciences, with minimum upper second class or equivalent; M.Sc with course work and thesis research in a subject relevant to the selected fellowship topic; relevant research experience (University or NARS staff eligible for study leave are preferred). Maximum age of 35 years; eligibility for registration at an accredited university.

Contacts: *ICIPE, Head, Capacity Building or from the Dept of Biology, P O Box 1176, Addis Ababa, Ethiopia or Head of Capacity Building, ICIPE, P O Box 30772, Nairobi, Kenya Fax: +254 2 86110/803360 http://www.icipe.org/arppis/*

African Wildlife Foundation: Charlotte and Macarthur Fellowship and Award

In 1996, the African Wildlife Foundation (AWF) introduced the Charlotte Conservation Fellowships to provide support for African nationals pursuing masters' degree courses or doctoral research. Charlotte Fellowships enable recipients to improve their qualifications, update their skills, acquire the latest information on natural resource management and adapt new technologies to their work. The Program is committed to enhancing the effectiveness and impact of African nationals in the field of conservation through the increased knowledge, skills, and credentials obtained through an advanced degree. AWF is offering scholarships for full or partial M.Sc or Ph.D. programs with field research components that produce knowledge offering solutions or insight into specific conservation challenges that complement AWF's conservation programs in the African Heartlands. Expenses that AWF expects to cover under the scholarship will vary depending on the individual program selected by the scholarship recipient.

Contacts: *Plot 27, Old Moshi Road P.O. Box 2658 ARUSHA, TANZANIA; Phone: +255 27 2509616; Fax: +255 27 2544453; africanwildlife@awf-tz.org*

26 Baker Street, South Entrance, Ground Floor, Rosebank 2196 JOHANNESBURG, SOUTH AFRICA; Phone: +27-11-447-2399; Fax: +27-11-447-7293; africanwildlife@awfsa.org

Blvd. du 30 juin No. 2515 Immeuble AFORIA (ex. Shell) 6eme Niveau, B.P. 2396 Kinshasa/Gombe REPUBLIQUE DEMOCRATIQUE DU CONGO; Phone: +243 (0) 8139-97586

Britak Centre Mara Ragati Road, P.O. Box 48177, 00100 NAIROBI, KENYA; Phone: +254 20 2710367; Fax: +254 20 2710372; africanwildlife@awfke.org

1400 Sixteenth Street, NW Suite 120 WASHINGTON, D.C. 20036, U.S.A; Phone: +1 202 939 3333; Toll free: +1 888 494 5354; Fax: +1 202 939 3332; africanwildlife@awf.org

50 Independence Avenue P.O. Box 50844 Ridgeway, LUSAKA, Zambia; Phone: + 260 1 257074; Fax: + 260 1 257098; africanwildlife@iwayafrica.com http://www.awf.org

African Women Leaders in Agriculture and the Environment

Grants to women from developing countries to pursue graduate studies in agriculture and the environment in the United States. Fellows enrolled in an African university pursue further course work and leadership training. Completion of studies and documentation of research will take place at the African university from which the degree will be awarded.

Contacts: *AWLAE, 1611 North Kent Street, Suite 600, Arlington, VA 22209, USA. Or The Information Officer, P. O. Box P.O. Box 60745, Nairobi, Kenya. Phone: +1(703) 525-9430; Fax: +1 (703) 525-1744 (USA) Phone: +254 2 711590/712966; Fax: +254 2 712839 (Kenya) http://www.winrock.org/leadership/files/Newsltr00FINAL.pdf*

African Women's Development Fund

The African Women's Development Fund (AWDF) funds local, national, sub-regional and regional organizations in Africa working towards women's empowerment. The AWDF is an institutional capacity-building and programme development fund, which aims to help build a culture of learning and partnerships within the African women's movement. In addition to awarding grants, the AWDF attempts to strengthen the organizational capacities of its grantees. The AWDF funds work in six thematic areas: (i) women's human rights, *(ii)* political participation, *(iii)* peace building, *(iv)* health, reproductive rights, *(v)* economic empowerment, (vi) HIV/AIDS. The AWDF makes grants in three cycles each year with applications accepted year round. Applying organizations should preferably have been in existence for three years and women's organizations have to be led and managed by women. Location: Africa.

Contacts: *The African Women's Development Fund; 25 Yiyiwa Street; Abelenkpe, Accra; P.M.B CT 89, Cantonments; Accra, Ghana*
Phone: +233 21 780467/7, Fax: + 233 21 782502
awdf@awdf.org or grants@awdf.org
http://www.awdf.org

AFRICAN-AMERICAN INSTITUTE

For African master's or doctoral students who wish to study in the U.S., in all areas of study, but especially agriculture, economics, public and business administration, engineering, health, and education. Applicants are selected by their respective countries and must be in residence in a Sub-Saharan country at the time of application. Deadline: varies by country (check website for details).

Contacts: *African-American Institute, Chanin Building, 380 Laxington Avenue, New York, NY 10168-4298 USA*
Phone +212 949 5666; Fax: +212 682-6174
aainy@aaionline.org
http://www.aaionline.org

ALCAN PRIZE FOR SUSTAINABILITY

One Alcan Prize of $1 million USD is awarded each year. Alcan Training Grants to the value of US$15,000 are awarded to the remaining nine short listed organizations, with the specific intention of investing in certifiable training and capacity building for the organization, e.g. Cambridge University Post Graduate Certificate in Cross Sector Partnership. The Prize is open to any not-for-profit, civil society or non-governmental organization that achieves a positive impact on economic, environmental and/or social development. Locations are unrestricted. The Alcan Prize is to recognize and reward organizations that: progress the principles of sustainability (environmental, social and economic impacts) and demonstrate impact; meet or address specific development needs (health, education, access to water, etc.); value innovation and partnership as a path to sustainable success; and Create opportunities for not-for-profit, civil society and nongovernmental organizations to secure additional funds and resources.

Contacts: *Manager; Alcan Prize for Sustainability,*
manager@alcanprizeforsustainability.com
http://www.alcanprizeforsustainability.com/home_en

ALLEN FOUNDATION INC.

The Allen Foundation makes grants to projects that benefit human nutrition in the areas of education, training, and research. The policies and priorities of the Allen Foundation are: to make grants to fund relevant nutritional research; to support programs for the education and training of mothers during pregnancy and after the birth of their children; to assist in the training of persons to work as educators and demonstrators of good nutritional practices; and to encourage the dissemination of information regarding healthful nutritional practices and habits. Proposals received after this date will be considered for review the following year. Non-profit organization with Internal Revenue Service certification of 501(c) 3 tax-exempt status is eligible.

Contacts: *Allen Foundation Inc.; P.O. Box 1606; Midland, MI 48641-1606 USA*
http://www.allenfoundation.org

ALO/USAID - PARTNERING WITH HIGHER EDUCATION FOR INTERNATIONAL DEVELOPMENT

The Association Liaison Office for University Cooperation in Development (ALO) and the United States Agency for International Development (USAID). Through the Institutional Partnerships program, ALO solicits proposals from U.S. colleges and universities to partner with higher education institutions in developing countries. These partnerships further USAID strategic objectives and assist both partnering institutions to apply their expertise collaboratively to global development challenges.

Contacts: *The Association Liaison Office for University Cooperation in Development, 1307 New York Avenue, NW Suite 500 Washington, DC 20005-4701*
Phone (202) 478-4700; Fax: (202) 478-4715
http://www.aascu.org/alo/

AMERICAN ANTIQUARIAN SOCIETY

Residency fellowships 3-12 months at American Antiquarian Society Library; Mellon Post-Dissertation Fellowship for preparation of dissertation for publication. Mellon Post-Doctoral Fellowship. Submission deadlines vary contact programme officers for additional information and deadlines.

Contacts: *185 Salisbury St. Room 100 Worcester, MA 01609-1634*
Phone: 508-755-5221; Fax: 508-754-9069
http://www.americanantiquarian.org

AMERICAN ASSOCIATION OF UNIVERSITY WOMEN FELLOWSHIP

The AAUW Educational Foundation is the largest source of funding exclusively for graduate women in the world, supports aspiring scholars around the globe, teachers and activists in local communities, women at critical stages of their careers, and those pursuing professions where women are underrepresented. Awards are variable with specific requirements. In particular, applicants must be citizens in a country other than the United States, or you must hold a non-immigrant visa if you reside in the United States, must complete an academic degree equivalent to a bachelor's degree from a U.S. college or university by the time of application, must intend to devote themselves full time to proposed academic plan during the fellowship year, must intend to return to their home countries to pursue a professional career. Awards are variable.

Contacts: *AAUW Educational Foundation 1111 Sixteenth St. NW Washington, DC 20036, USA;*
Phone: +1(202)7287602;
foundation@aauw.org
http://www.aauw.org

Aspen Institute

Please check the website for detailed information regarding deadlines and eligibility requirements.

Contacts: *info@aspeninstitute.org*
http://www.aspeninst.org

Atlantic Salmon Federation Fellowships

ASF Fellowships are offered annually to individuals seeking to improve their knowledge or skills in advanced fields while looking for solutions to current problems in Atlantic salmon biology, management and conservation. The Fellowships may be applied toward a wide range of endeavors including salmon management, graduate study, and research. Applicants need not be enrolled in a degree program but must be legal residents of the United States or Canada.

Contacts: *http://www.asf.ca/Awards/OlinApp.html*

Australian Development Scholarships Programme

Australian Development Scholarships (ADS) provide opportunities for people from selected developing countries to undertake post-secondary level study in Australia. These Scholarships allow people to gain knowledge and skills which will help the development of their home countries when they return home after finishing their studies. Australian Development Scholarship students study a wide range of disciplines at Australian universities and most technical training institutions. The scholarship numbers and fields of study for each country are determined annually as part of Australia's bilateral development assistance program with that country. Applicants must be citizens of eligible developing countries.

Contacts: *GPO Box 887; Canberra ACT 2601, Australia;*

Phone: 61.2.6206.4000; Fax: 61.2.6206.4880;
http://www.ausaid.gov.au/scholar/studyin.cfm

BILL AND MELINDA GATES FOUNDATION: THE GATES MILLENNIUM SCHOLARS PROGRAMME

The Gates Millennium Scholars (GMS), funded by a grant from the Bill and Melinda Gates Foundation, was established in 1999 to provide outstanding African American, American Indian/Alaska Natives, Asian Pacific Islander Americans, and Hispanic American students with an opportunity to complete an undergraduate education for those students pursuing studies in mathematics, science, engineering, education or library science. Awards are available to high school seniors, college freshman, sophomores, juniors, and seniors, and graduate students.

Contacts: *Bill and Melinda Gates Foundation; PO Box 23350; Seattle, WA 98102 USA, Phone: (206) 709-3100 (Reception); (206) 709-3140 (Grant Inquiries), info@gatesfoundation.org;*
http://www.gatesfoundation.org/education/scholarships

BLACKSMITH INSTITUTE

Blacksmith develops and implements solutions for pollution-related problems in the developing world. They work cooperatively with partnerships of donors, governments, NGO's and others, and provide strategic, technical, and financial support to local champions as they strive to solve specific, pollution-related problems in their communities. In 2003, Blacksmith Institute launched the Polluted Places Initiative to address severely polluted sites throughout the world. By identifying polluted sites directly through an on-line nomination process, researching each site and publishing dossiers online, and conducting site assessments to consult with local stakeholders. Blacksmith Institute also works with governments and local NGOs to raise awareness about pollution, create a sound knowledge base on environmental quality through monitoring and research, and strengthen legislation and legal frameworks to curb polluting practices by offering different types of support.

Contacts: *Blacksmith Institute; 2014 5th Avenue; New York, NY 10035 USA*
applications@blacksmithinstitute.org
http://www.blacksmithinstitute.org

BRONX ZOO TEACHING FELLOWSHIPS

The Education Department offers spring, summer, and fall fellowships to qualified college students, recent graduates or graduate students. Spring and fall fellows team-teach in our onsite school programs for elementary aged children. Spring and fall fellows must be able to work 20 hours a week (mornings) for approximately 18 weeks. Summer fellows work in the Zoo's camp programs. Summer fellows must be able to work 35 hours a week for 11 weeks. Experience in teaching is highly desirable. An interest in wildlife conservation and the ability to relate to children are essential. Field of study are zoology, biology, ecology, and education.

Contacts: *http://www.bronxzoo.com*

BUDAPEST OPEN ACCESS INITIATIVE

To support open access to international research and scholarship, OSI provides funding to support 50 Public Library of Science Institutional memberships for institutions in developing and transition countries. Authors from the least developed countries receive a waiver of the publication charge upon request, The institutional memberships provides the following benefits to the recipient institutions: the right for any of the institution's research staff or students to automatically qualify for a publication-charge waiver when their manuscripts are submitted to any of the Public Library of Science journals is accepted for publication; the right to list all the relevant publications generated at the institution on a special page on the Public Library of Science web site, customized with the institution's name. The institutional membership grants will be for a period of twelve months with decisions on funding made on a quarterly basis. Institutions must be engaged in research in biological and/or medical sciences. A geographical balance is sought in funding but generally funding is open to all developing countries

Contacts: *http://www.soros.org/openaccess/index.shtml*

CAMILLE AND HENRY DREYFUS FOUNDATION

The program invites experts in environmental science to submit proposals for the training of Ph.D. chemists and chemical engineers in research activities related to chemistry and the environment. Each awardee may then appoint a postdoctoral scientist to spend up to two years in a program intended to stimulate the visiting scientist to carry out further research in support of environmental understanding and values. The program is open to all academic and other not-for-profit organizations that have well-established research efforts in environmental science or engineering. These research activities need not be located in traditional departments in the chemical sciences. Examples include but are not limited to fundamental science or engineering in topics related to the troposphere or stratosphere, aquatic or marine settings, soil or groundwater problems, environmental biochemistry, chemically benign synthesis and processing, or pollution control.

Contacts: *http://www.dreyfus.org/ep.shtml*

CANON NATIONAL PARKS SCIENCE SCHOLARS PROGRAMME

The program is a collaboration between Canon U.S.A., Inc., the American Association for the Advancement of Science and the US National Park Service. Thanks to a generous commitment by Canon U.S.A., Inc., the programme awards eight scholarships to Ph.D.

students throughout the Americas to conduct research critical to conserving the national parks of the region. Research projects in the biological, physical, social and cultural sciences are eligible, as well as projects in a new category — technology innovation in support of conservation science.

Contacts: *http://www.nature.nps.gov/canonscholarships/*

Centre for Environment and Development Studies in Cameroon

The Centre offers short training courses in environmental sciences, PhD-research, and development M.Sc curriculum "Environmental Science", conferences, consultancies and training of NGO-staff.

Contacts: *CEDC, PO BOX 410, Maroua, Cameroon*
Phone: + 237 293 061 or + 237292 950; Fax: + 237 293 391

Centre for Environmental Economics and Policy in Africa Research Grants

CEEPA aims to enhance the capacity of African researchers to conduct environmental economics and policy inquiry of relevance to African problems and increase the awareness of environmental and economic managers and policy makers of the role of environmental economics in sustainable development. With funding from the Swedish International Development Cooperation Agency (Sida), CEEPA is implementing the first phase of a regional program to strengthen research capacity in environmental economics and policy in Africa. Under this programme, research Grants are awarded on a competitive basis to research projects annually to increase capacity in research in environmental economics.

Contacts: *CEEPA Department of Agricultural Economics, Extension and*
Rural Development
Faculty of Natural and Agricultural Sciences
University of Pretoria; Pretoria 0002, Republic of South Africa
http://www.ceepa.co.za

Centre for Environmental Management and Planning

The Centre offers Training programmes with most courses charged at the UK international students' rates

Contacts: *CEMP, CORDAH, Kettock Lodge, Aberdeen Science and Technology Park*
Bridge of Don, Aberdeen AB22 8GU, UK
Phone: +44 1224 414 311/312; Fax: + 44 1224 414 250

Chesapeake Research Consortium Environmental Management Fellowships

This organization provides fellowships for professional development in the field of environmental management and protection. Fellows work in conjunction with the Chesapeake Bay Program for 1 to 2 years. Fellows are recommended to have degrees in natural science or environmental management, or comparable work experience. Fellowships become available on a year-round basis. A two-year programme in which participants work primarily as the administrative support for a variety of subcommittees and workgroups of the Chesapeake Bay Program in Annapolis, MD. During this two-year appointment, Fellows have the opportunity to learn about and take part in the shaping of environmental policy. Priority is given to candidates with degrees in natural sciences or environmental policy and planning. Strong writing, communication, and general computer skills are desirable. Some assignments may require more specialized education and experience, such as computer and statistical skills or agricultural education/experience. A competitive vacation and benefits package is included.

Contacts: *http://www.chesapeake.org/crc/fellow.html*

Chevening Scholarships the British Council

Full scholarships for an academic year and short vocational course (minimum length: 3 months). Three types of scholarships: Type A covers tuition fees only, Type B are full scholarships that cover tuition fees, monthly stipend and various one-off allowances. Type C vary from award to award and cover part/full stipend and/or allowances. Standard requirements include: graduate with proven academic skills and having an admission to a UK Higher Education Institution (HEI); young professionals in their early to mid career (under 35 years old); committed to return to their country and contribute to the socio-economic development of their country through implementing the new skills and knowledge acquired in the UK; established in a career, with a track record of excellence and achievement, and the prospect of becoming a leader in his/her chosen field; have a clear idea how their scholarship will benefit their country on their return; have good English Language skills (IELTS of 6.5 minimum); PhD's and M.Phil's applicants are considered when there is a co-funding partner. Deadlines vary according to countries

Contacts: *http://www.chevening.com/about/scholarships*

CIDA - Marine Scholarship and Environment Programme

This programme supports individual mid-career professionals in developing countries who are working in the field of marine affairs with scholarships, and requires applicants to be professionals wanting to pursue studies at the Master's level; currently working in marine affairs; originate from a coastal country eligible for Canadian official development assistance; ensure the proposed area of study corresponds with your work; and have the support of their employers. Check website for additional information and application processes.

Contacts: *200 Promenade du portage Hull, Quebec KIA OGA Canada;*

Phone: +1 (819) 997-2489.
http://www.acdi-cida.gc.ca

COLUMBIA UNIVERSITY FELLOWSHIP: COLLEGE OF ARTS AND SCIENCES

All applicants to Columbia's Graduate School of Arts and Sciences are considered for University fellowship at the time of application for admission, provided the application is complete by the stated deadline. All continuing GSAS students must apply for renewal of their fellowships or for fellowship consideration if they did not receive an award the prior year. Each individual department has procedures and deadlines for fellowship consideration.

Contacts: *http://www.columbia.edu*

COMMONWEALTH FOUNDATION

The Foundation provides grants that enable people from the Commonwealth to participate in activities such as short training courses, workshops, seminars, conferences, cultural festivals, exchanges and study visits in other Commonwealth countries. The Commonwealth Foundation's projects are divided into three areas: governance and democracy; sustainable development; and culture and diversity. The Foundation gives priority to activities where the direct beneficiaries are women and men who: are young (35 years old or under) or mid-career staff; are working in civil society organizations directly engaged in development and/or cultural work; are employed in work which has direct relevance to the subject of the activity; are in a position to transfer their learning to others. Grants are awarded on a quarterly basis. Priority is given to activities taking place in developing countries which are members of the Commonwealth Foundation.

Contacts: *Commonwealth Foundation; Marlborough House; Pall Mall,*
London SW1Y 5HY United Kingdom;
Phone: +44 (0)20 7930 3783; Fax +44 (0)20 7839 8157
http://www.commonwealthfoundation.com

COMMONWEALTH SCHOLARSHIP AND FELLOWSHIP PLAN

These scholarships assist Commonwealth students and academics of high intellectual caliber to study in other commonwealth countries so as to strengthen common ideas to promote the equality of educational opportunity at the highest level. Subjects offered include: agriculture, commerce and management, computer science, engineering, languages, medical sciences, natural sciences, social sciences. Applicants must have Masters degree/equivalent in the proposed field of study within the last ten years; First Division throughout academic career; at least three years work experience in relevant discipline; Maximum age limit 34 years.

Contacts: *http://www.britishcouncil.org.pk/education/paklcbc.htm*

Commonwealth of Learning

The Commonwealth Of Learning-PROTEIN program aims to support innovative uses of open and distance learning (ODL) to address poverty alleviation in rural and peri-urban areas. In particular, COL-PROTEIN is seeking to support initiatives that will adopt open and distance learning and information and communication technologies (ICT) to help build capacity. Proposals that demonstrate effective collaboration and networking among organizations/institutions are especially welcome. Proposals are invited from qualified applicants in developing Commonwealth countries from Africa, Asia, the Pacific and the Caribbean.

Contacts: *Commonwealth of Learning; Attention: COL-PROTEIN; 1055 West Hastings St., Suite 1200; Vancouver, BC V6E 2E9 Canada*
Fax: +1 604 775 8210
protein@col.org
http://www.col.org

Darwin Researcher Exchange Programme Awards

The RXP exchange awards fund short term exchange visits for postdoctoral researchers to make international connections. This can be a single visit between two weeks and three months duration, or multiple visits. The award is to the individual, not the link and cannot exceed £5,000. The awards are for travel and subsistence. A small amount of the award can also be used for consumable equipment, up to £250, although it is expected that the receiving lab will provide most of these resources. Those requesting an amount for consumable equipment should provide a breakdown of items to be purchased. It cannot be used to cover bench fees, utility bills or staffing overheads, non-consumable equipment or salaries. Researchers can apply for an award for one or more visits. These should be between two weeks and three months in total.

Contacts: *http://www.britishcouncil.org/science-rxp*

Department of Energy: The Hispanic Scholarship

Since 1990 the Department of Energy and, more specifically, the Office of Environmental Management, has invested in the talents of promising Latino students through the Environmental Management Scholarship. The programme provides scholarship to 4-year university students who are pursuing majors pertinent to the Department of Energy's goal of environmental restoration and waste management.

Contacts: *http://www.hsfi .org/scholarships/energy.asp*

DISCOVERY CHANNEL GLOBAL EDUCATION PARTNERSHIP

Discovery Channel Global Education Partnership is a nonprofit, charitable organization dedicated to reaching across the global information divide with the tools and training necessary to extend the power of technology and information to under-resourced communities around the world. The Discovery Channel Global Education Partnership's Learning Centers are versatile community resources, designed to benefit children, teachers, parents, local HIV/AIDS and health care organizations, community development groups, and others who desire to receive or distribute information via television and video. Learning Centers are established in existing structures that can serve as centers for community learning. A Learning Center may be located in a school or community center, a clinic, library, mobile van, or even a donkey-drawn TV cart. Through local staff trained by the Global Education Partnership, teachers at each Learning Center receive three years of training and monitoring in the use of video as a teaching tool, as well as support in utilizing the Learning Center as a local education and information hub.

Contacts: *Discovery Channel Global Education Partnership, One Discovery Place, Silver Spring, MD 20910, USA.*
Phone: +1.240.662.2986; Fax: +1.240.662.1847;
partnership_info@discoveryglobaled.org
http://www.discoveryglobaled.org

DOS BUREAU OF EDUCATIONAL AND CULTURAL AFFAIRS – FULBRIGHT INTERNATIONAL EXCHANGE

The flagship international educational program sponsored by the United States Government, widely known as the Fulbright Programme, is designed to "increase mutual understanding between the people of the United States and the people of other countries..." With this goal as a starting point, the Fulbright Programme has provided more than 230,000 participants - chosen for their leadership potential - with the opportunity to observe each others' political, economic and cultural institutions, exchange ideas, and embark on joint ventures of importance to the general welfare of the world's inhabitants. Check website for additional information.

Contacts: *http://www.iie.org/cies/us_scholars*

ECHOING GREEN FOUNDATION

Echoing Green awards two-year fellowships to emerging social innovators. Annually, Fellowships are awarded to individuals with innovative ideas for creating new models for tackling seemingly unsolvable social challenges. These fellowships offer them the opportunity to develop and test their ideas. This is not a scholarship programme. Fellows do not develop their ideas in an academic setting; they work in the community. They launch, manage and grow organizations that implement and continually expand their ideas for creating lasting social change. Applicants should be emerging social entrepreneurs whose plan will result in a sustainable organization. Applicants of all nationalities, working in any country are eligible to apply for the fellowship.

Contacts: *Echoing Green; 494 Eighth Ave; 2nd Floor; New York, NY 10001 USA*
Phone: 212-689-1165; Fax: 212-689-9010
info@echoinggreen.org
http://www.echoinggreen.org

Ford Foundation

i) *International fellowship programme:* The International Fellowships Programme provides support for up to three years of formal graduate-level study leading to a masters or doctoral degree. Fellows are selected from countries in Asia, Africa, the Middle East, Latin America, and Russia, where the Ford Foundation maintains active overseas programs. U.S. nationals are not eligible, although Fellows may study in the US. The foundation has invested $280 million through 2010.

ii) *postdoctoral fellowships for minorities:* Limited to U.S. citizens in one of the following ethnic minority groups: Native Americans, Alaskan Natives, African Americans, Mexican Americans, Native Pacific Islanders, and Puerto Ricans. Only individuals engaged in a teaching and research career or those planning such a career are eligible for this program. Previous Ford Foundation Postdoctoral Fellow may not reapply. The Foundation provides approximately 20 fellowships annually in the following disciplines: behavioral and social sciences; humanities; engineering; mathematics; physical sciences; life sciences; interdisciplinary programs comprised of two or more eligible disciplines;

iii) *leadership for a changing world:* The programme seeks to recognize, strengthen, and support leaders, and to highlight the importance of leadership in improving lives, and to confirm that resourceful leaders are bringing about positive change in American communities. Each year, the program recognizes twenty leaders and leadership groups not broadly known beyond their immediate community or field. These outstanding leaders and leadership groups work in such areas as economic and community development, human rights, the arts, education, human development, sexual and reproductive health, religion, media, and the environment.

Contacts: *http://www.fordfound.org/what/initiatives.cfm*
http://www.aau.org
http://www.fordifp.net
http://www.iucea.org
http://www.nationalacademies.org/fellowships/fordpost.html
http://www.fordfound.org

Fulbright Grants

The Fulbright Program offers a variety of grant opportunities, including comprehensive Fulbright grants and Fulbright travel grants. Additional grants are available on a country-by-country basis, including opportunities in teaching and business. Length of grants: Most of the grants described will be awarded for programs of study or research that will require one academic year. Specific academic qualifications are contained in the Regional Summaries or Individual Country Summaries. Applicants must be U.S. citizens at the time of application, and must hold a B.A. degree or the equivalent before the beginning date of the grant but a doctoral degree at the time of application, unless otherwise noted.

Applicants must have sufficient proficiency in the written and spoken language of the host country. *(i) Foreign student programme:* Full grants for graduate study at U.S. institutions including travel costs. All grantees are required to be on "J" visas under Exchange-Visitor Program No. G-1-1. Please contact the U.S. Embassy in your country for country-specific requirements. Deadlines are variable depending on countries of applicants; *(ii) Teacher and administrator exchange program administered by graduate school:* Fulbright Teacher and Administrator Exchange Programme – Participate in an academic year or one semester direct exchange of positions with teachers from other countries. Participate in a six-week administrative exchange.

Contacts: *809 United Nations Plaza, New York, N.Y. 10017, USA*
Phone: Tel: +1 (212) 984-5583
http://www.iie.org

Garden Club of America Scholarships

The Garden Club offers a number of awards for work in horticulture, landscape architecture, agriculture, tropical botany, and tropical forest preservation and environmental studies. Fellowships are open to US and non-US citizens.

Contacts: *http://www.gcamerica.org/scholarships.php3*

George Wright Student Travel Scholarships

The George Wright Society created the George Melendez Wright Student Travel Scholarship for the George Wright Society Conference. The Society seeks to encourage students from diverse ethnic and cultural backgrounds to participate in its biennial conference and to develop an interest in the conservation and preservation of parks, protected areas, and cultural sites throughout the world. It is the Society's hope that by offering these scholarships, students will be encouraged to pursue a profession in fields directly related to parks, protected areas, and cultural sites.

Scholarship recipients have the opportunity to participate in conference educational sessions and learning activities, as well as to network with peers and professionals from a wide variety of fields related to protected areas conservation.

Contacts: *http://www.georgewright.org*

German Academic Exchange: Funding for Academic Cooperation and Exchange

The German Academic Exchange Service (DAAD) is a joint organization of Germany's higher education institutions and is responsible for promoting international academic relations, primarily through the exchange of students, academics and researchers. The mandate of the predominantly publicly funded DAAD is to raise the international profile of Germany's higher education institutions and to serve as a "mediating organization" in the government's foreign, European, development and higher education policies. DAAD's main objectives include: (i) encouraging outstanding young students and academics from abroad to come to Germany for study or research visits and, if possible, maintaining contact with them as life-long partners, (ii) qualifying young German researchers and professionals at the very best institutions around the world in a spirit of tolerance and openness, (iii) promoting the internationality and appeal of Germany's universities and higher education institutions, (iv) maintaining or establishing German studies including German language, literature and cultural studies at an appropriate level at major foreign universities, and (v) helping the developing countries of the South and the reforming states of the East to establish efficient higher education structures. Several funding options exist for both Individual and Group projects. Funds are available for Germans going to Africa, Africans going to Germany, and Africans to work in Africa. See website for more detailed information.

Contacts: *DAAD Nairobi; P.O Box 14050-00800; Nairobi, Kenya*
Phone: +254-20-2729741/+254-20-2722660/ +254-20-2722662;
Fax: +254-20-2716710
Email: info@daadafrica.org
http://www.nairobi.daad.de

Gladys Krieble Delmas Foundation

Gladys Krieble Delmas Foundation offers both protectoral and postdoctoral grants for between $500 and $14,500 for travel and research. Funds may be available for publication of research done under the grant.

Contacts: *521 5th Avenue, St. 1612, New York, NY 10175-1699*
Phone: 212-687-0011; Fax: 212-687-8877
delmasfdtn@aol.com
http://www.delmas.org

Hubert H. Humphrey Fellowship Programme

This fellowship Programme brings accomplished professionals to the United States at a midpoint in their careers for a year of study and related professional experiences. During their Fellowship year, Humphrey Fellows participate in programmes that combine graduate-level academic coursework with professional development activities. The programme is not a degree programme, but rather is designed to provide broad professional enrichment through a combination of activities tailored to each Fellow's interests. Humphrey Fellowships are limited to one academic year, preceded, if appropriate, by a period of English language training. Of the many eligible fields, natural resources and environmental management is presented on the top of substantially long list.

Contacts: *Public Affairs Section of the U.S. Embassy or the Fulbright Commission*
http://www.iie.org

INSTITUTE OF ECOLOGY AND RESOURCE MANAGEMENT IN SOUTHERN AFRICAN

Advance Certificate Course in Wildlife Management for Protected Areas (40 modules).

Contacts: *PO Box X3015, Hoedspruit, 1380, South Africa; Phone: +27 15 793 2621; Fax: +27 15 793 2621*

INTENSIVE TRAINING PROGRAMME ON ENVIRONMENTAL MANAGEMENT - WEST AFRICA

Short Course for 165 hours (5 modules on Sustainable Development and Environmental Management Systems (30 hours); Environmental Care Systems (30 hours); Eco-design and Eco-labeling (30 hours); Environmental taxes (30 hours); Environmental Responsibility in World Trade (45 hours).

Contacts: *International Centre for Enterprise and Sustainable Development*
P. O. Box 16461, Airport-Accra, GHANA
Programme Co-Director Human Ecology Department Vrije Universiteit
Brussels Laarbeeklaan 103, B 1090 Brussels, BELGIUM

INTERNATIONAL TROPICAL TIMBER ORGANIZATION FREEZAILAH FELLOWSHIP FUND

ITTO offers fellowships through the Freezailah Fellowship Fund to promote human resource development and to strengthen professional expertise in member countries in tropical forestry and related disciplines. The goal is to promote the sustainable management of tropical forests, the efficient use and processing of tropical timber, and better economic information about the international trade in tropical timber.

Contacts: *http://www.itto.or.jp/*

ISABELLA FREEDMAN JEWISH RETREAT CENTER: ADAMAH FELLOWSHIP

A three or six month leadership training program for Jewish young adults — ages 20–29 — that integrates organic farming, sustainable living, Jewish learning, teaching, and contemplative spiritual practice. Fellows develop an intimate connection to the New England landscape and its agricultural rhythms. They leave the programme with both practical farming skills and a deep visceral understanding of how the combination of soil, sun, water, nutrients, and hard labor conspire to form the sustenance of our lives. They also learn how current modern agricultural practices undermine our planet's long-term ability to support life. In addition to agriculture, Fellows are given the opportunity to learn many other aspects of sustainability in elective courses. Specific courses depend on faculty expertise and Fellow interest in any given ADAMAH season. Fellows receive full room, board, limited health care, and a stipend which is paid at the conclusion of the Fellowship season. Fellowship is open to Jewish young adults.

Contacts: *http://www.isabellafreedman.org*

Japan Capacity Building Programme for African Agricultural Researchers

The capacity building program is sponsored by the Ministry of Agriculture, Forestry and Fisheries of Japan, and executed by the International Cooperation Center for Agricultural Education (ICCAE), Nagoya University. The purpose is to strengthen collaboration between Japan and agricultural research organizations whose mission is African development, and to deliver Japanese know-how of agricultural research methods and technologies to Africa. Training programs for promising young African researchers are offered in agricultural research organizations, preferably in Africa.

Contacts: *International Cooperation Center for Agricultural Education (ICCAE)*
Nagoya University; Furo-cho, Chikusa, Nagoya, 464-8601 JAPAN
Phone: +81-52-789-4225, Fax: +81-52-789-4222
http://www.iccae.agr.nagoya-u.ac.jp

Japan International Volunteer Center

Based on the concept of 'Resident-Lead Village Development', JVC engages in various activities including natural farming, mutual aid and environmental protection. JVC emphasizes the importance of 'life in harmony with nature and respect for local culture'. By promoting the restoration and preservation of the natural environment, mutual aid and people's existence in accordance with nature, it wants to enable the people to lead permanently stable lives in their villages and rural areas.

Contacts: *Marukou-Building 6F, 1-20-6, Higashiueno, Taitou, 110-8605 Tokyo – Japan*
Phone: +81-3-3834.2388; Fax: +81-3-3835.0519
http://www.jca.apc.org/jvc

John Deere Foundation

The John Deere Foundation, founded in 1948, primarily supports programs in education, health and human services, community development, and arts and culture. A Foundation Board of Directors meets at least once a year to award grants based upon: availability of funds, effectiveness of requestor's program, impact upon John Deere employees, and demonstrated level of community support.

Contacts: *John Deere Foundation, Deere and Company; One John Deere Place; Moline, IL 61265 USA*
Phone: (309)748-7955
http://www.deere.com

Johnson Research Assistantships

The Arthur S. and Helen May Johnson Research Assistantship is a non-renewable $4,000 assistantship for female engineering students. The assistantships were created to enable female engineering students to learn more about specific areas of technology and science and to acquaint them with the graduate student experience. Assistantship recipients are expected to work on a research project with a faculty member.

Contacts: *http://www.engr.ku.edu/scholarships/*

Joint Japan/World Bank Graduate Scholarship Programme

Submit evidence of current unconditional admission (copies of formal acceptance letters) to at least one university and proof of application to another university located in any World Bank member country except the applicant's country. Be a national of a World Bank member country eligible to borrow. Be under the age of 45, and preferably under 35 years of age Have at least 2, preferably 4 to 5, years of recent full-time professional experience after a university degree in the home country or in another developing country and usually in public sector. Hold a bachelor's degree or its equivalent in a development-related field; have a superior record in previous academic work. Propose a program of study related to development at the master's level.

Contacts: *The World Bank, Joint Japan/World Bank Graduate Scholarship Program*
1818 H Street, NW - MSN J4-402, Washington, DC 20433 USA
Phone: +1 (202) 473-6849; Fax: +1 (202) 522-4036
jjwbgsp@worldbank.org; http://www.worldbank.org

Joseph L. Fisher Doctoral Dissertation Fellowships, Resources for the Future

This doctoral dissertation fellowship carries a stipend of $12,000. Proposals from the physical or biological sciences must have an immediate and obvious link to environmental policy matters to be considered. Graduate students in the final year of their dissertation research. Fellowship candidates must have completed the preliminary examinations for the doctorate prior to the application.

Contacts: *Coordinator for Academic Programs -Resources for the Future*
1616 P Street, NW, Washington, DC 20036-1400 USA
Phone: +1 (202) 328-5060, Fax: faxed applications not accepted
http://www.rff.org

Louis Leakey Foundation

The Louis Leakey Foundation offers the Franklin Mosher Baldwin Memorial Fellowships, which strive to usher in a new era in African paleoanthropology. This fellowship is intended for scholars with citizenship in an African country who seek to obtain an advanced degree or specialized training in an area of study related to human origins research. The award is for a program of approved special training and/or advanced training towards an MA, PhD, or equivalent and is limited to two years of support. Refer to website for application procedures.

Contacts: *http://www.leakeyfoundation.org*

Mote Marine Laboratory College Intern Program Research Programmes

Mote offers a variety of internship opportunities for undergraduate/graduate students and recent college graduates interested in a career in the ocean sciences. Mote internships are designed to inspire students with a life-long appreciation and commitment to the conservation and sustainability of our oceans by providing unique, hands-on opportunities to conduct ocean science research, education or support services under the mentorship of Mote scientists, educators and animal care specialists.

Contacts: *Mote Marine Laboratory 1600 Ken Thompson Parkway Sarasota,*
FL 34236, United States
Phone: +1 (941) 388-4441; Fax: +1 (941) 388-4312
intern@mote.org
http://www.mote.org

Mountain Institute

The Mountain Institute is a non-profit scientific and educational organization committed to the preservation of mountain environments and advancement of mountain cultures around the world. TMI focuses its work in mountain areas and on mountain people. It has offices and community-based programs in the Andean, Appalachian, and Himalayan mountain ranges collaborating with local people to strengthen their communities and to conserve their natural resources and cultural heritage. TMI works to connect isolated mountain people world wide through the Mountain Forum (MF)

Contacts: *1828 L Street NW, Suite 725, DC 20036 Washington – USA*
Phone: +1-202-452.16.36; Fax: +1-202-452.1635
http://www.mountain.org

Open Society Institute

The Open Society Institute (OSI) established the Community Fellowships Program to encourage and support community leaders creating innovative public interest projects that address critical social issues. The goal of the fellowships program is to provide individuals an opportunity to apply their abilities and vision in community-led projects that empower and improve the quality of public life. The Community Fellowships Program identifies and supports social entrepreneurs, activists and organizers who engage their passion, ingenuity and dynamism to remove social barriers and create new opportunities for disadvantaged communities and individuals. Community Fellows infuse a wide range of social change tools including, but not limited to, organizing, service delivery, and advocacy to address such community-based issues as youth development, the environment, workers' rights, the digital divide, education, and criminal justice.

Contacts: *Open Society Institute-New York 400 West 59th Street; New York, NY 10019*
Phone: (212) 548 0600; Fax :(212) 548 4679; ttp://www.soros.org

Organization for Tropical Studies: Research Fellowships

OTS offers research fellowships to graduate students enrolled in degree programs at OTS member institutions and to OTS alumni. Post-course awards are small amounts for up to two weeks of study after a course. Research Fellowships are to assist thesis research in tropical biology and related fields – proposals for research at OTS biological stations receive priority.

Contacts: *http://www.ots.duke.edu/index_nao.php*

Organization of Petroleum Exporting Countries Fund for International Development

Scholarship Award: These scholarships are awarded to one potential candidates who meet all of the following requirements: be a national of a developing country; be a mid-career professional with a Bachelor degree; has obtained admission to a Master's program, or its equivalent, of up to two years from an accredited academic institution recognized by the applicant's country; and be committed to return to his/her home country upon completing the program.

Contacts: *OPEC Fund for International Development*
Parkring 8, A-1010, Vienna, Austria
info@ofid.org; http://www.opecfund.org

OVERSEAS RESEARCH STUDENTS AWARDS SCHEME

The Overseas Research Students Awards Scheme (ORS) grants awards on a competitive basis to international postgraduate research students of outstanding merit and research potential. The award scheme is administered by Universities UK on behalf of the Department for Education and Skills (DfES), and is reimbursed by the DfES through the UK funding councils. These awards are available to students of any nationality who are accepted for registration for the MPhil/PhD degree. They are competitive and depend on academic merit and research potential. Check website for a full list of participating institutions.

Contacts: *http://www.universitiesuk.ac.uk*

PERCY FITZPATRICK INSTITUTE, UNIVERSITY OF CAPE TOWN

M.Sc Course in Conservation Biology; One year course: Applicants with at least honours or an equivalent four-year degree are invited to apply for a one-year Master of Science (M.Sc) course in conservation biology. The medium of instruction and examination at UCT is English and proficiency in English is therefore a prerequisite (proofs of TOEFL or PTEEP certificates or other good English qualifications required); A one or two page summary on how one's career vision requires the M.Sc. training; completed UCT Admission Forms (with the admission fees). Applications must reach the FitzPatrick Institute by the end of August to be considered for the following year.

Contacts: *The Director, FitzPatrick Institute*
(University of Cape Town) Rondebosch 7701, South Africa
Phone: +27 21 650 3290/1; Fax: +27 21 650 3295;
http://www.fitzpatrick.uct.ac.za

PEW FELLOWS PROGRAMME IN MARINE CONSERVATION

The Pew Fellows program began in 1990 as the Pew Scholars Program in Conservation and the Environment, to fund primarily academics to address a broad range of biodiversity issues. In 1994 the program was expanded beyond academia and recipients became known as Pew Fellows. In 1996 the program focus was narrowed to marine conservation exclusively. The PFP now awards mid-career fellowships that support individuals/teams to engage in policy-focused research, applied science, and outreach for marine conservation and challenges arising from land-sea connections.

Contacts: *http://www.pewoceanscience.org/fellowssite/fellows.php*

POPULATION COUNCIL FELLOWSHIPS

Offers to support work leading to a doctoral degree, postdoctoral research, or mid-career. Although aimed primarily for supporting training in the population field, funding is also provided for social science discipline such as economics, sociology, anthropology or geography. Awards are made only to applicants whose proposals deal with the developing world.

Contacts: F*ellowship Co-ordinator, Policy Research Division*
Population Council, One Dag Hammarskjold Plaza, New York, NY 10017, USA
Phone: +1(212) 339-0671; Fax: +1 (212) 755-6052
http://www.popcouncil.org

ROGER REVELLE FELLOWSHIP IN GLOBAL STEWARDSHIP

The fellowship is awarded to an emerging leader in the scientific community who shows an ability to make a significant contribution to domestic or international environmental issues, encompassed under the umbrella of global stewardship. The focus of the fellowship is on human interaction with ecosystems, which may include work in such areas as population, sustainable development, global climate change, food security, and related environmental concerns. Applicants must be U.S. citizens and must have a Ph.D. or an equivalent doctoral-level degree by the application deadline in any physical, biological or social science, or any relevant interdisciplinary field, and at least three years of post-degree professional experience.

Contacts: *http://www.fellowships.aaas.org/revelle*

RUSSELL E. TRAIN FELLOWSHIPS

Through Russell E. Train Fellowships, Education for Nature Program invests in the academic training of conservationists in Africa, Asia, and Latin America in a wide variety of disciplines so that they may gain the knowledge and skills necessary to manage natural resources in complex contexts. To be eligible for a fellowship, applicants must be a citizen or legal resident of a participating country and must have a minimum of two year's work experience in conservation. Applicants must be enrolled, admitted, or have applied to an institution of higher education and must begin their studies within one year from the application deadline for their respective countries.

Contacts: *http://www.wwf-efn.org/fellowships.cfm*

SOCIETY OF ENVIRONMENTAL TOXICOLOGY AND CHEMISTRY: ENVIRONMENTAL EDUCATION

SETAC offers Pre-Doctoral Fellowship Awards and Travel Grants. Pre-Doctoral awards are available to full-time graduate students doing research in the areas of environmental toxicology, environmental chemistry, hazard assessment or risk analysis. Applications must be submitted by the dissertation director, who must be a member of SETAC.

Contacts: *http://www.setac.org*

Stichting School van Z.M. Koning Willem III

The Foundation provides financial support to projects that promote knowledge of forestry, the training of employees, teachers or information officers, the creation or restoration of training, information or educational centres, the production of information or educational material, and research aimed at the preservation, maintenance, or restoration of forests. The Foundation supports the use, protection, or restoration of forests and forestry woody and herbaceous vegetation occurring under more or less natural conditions in association with fauna and abiotic factors. It also supports projects that look at the use of forest for commercial, educational and recreational purposes; the maintenance of forest for its value to nature and the environment, the use of products originating from the forest, and the use of substitute products in the interest of the forest.

Contacts: *Koninklijk Park 1, 7315 JA Apeldoorn - the Netherlands*
Phone: +31-55-521.97.09;Fax: +31-55-5224462
http://www.koningsschool.nl

Trent R. Dames and William W. Moore Fellowship

Grants are to provide for the exploration of new applications of geotechnical engineering or the earth sciences to social, economic, environmental and political issues. Applicants must be practicing engineers or earth scientists, professors or graduate students. Membership in the Society is not a requirement. Selection is based on technical and social value of the proposed research, ability of the research to advance the science and profession of engineering and to enhance the overall welfare of mankind, relevant professional experience of the applicant, and applicant's prior contribution to the engineering profession and/or the applied earth sciences. Award is used for expenses for research and experiments as stated in the eligibility requirements. Check website for additional information.

Contacts: *http://www.asce.org*

Teresa and H. John Heinz III Foundation Scholarships

The program provides enhancement support for doctoral dissertation and master's thesis (or project) research. A total of 16 one-time, one-year awards are offered annually. Funded research must have public policy relevance that increases society's understanding of environmental concerns and proposed solutions. Individuals with dissertations or theses or projects that do not meet these criteria will not be considered. Recipients must participate in either a doctoral or master's program at the following institutions: Carnegie Mellon University, Cornell University, Harvard University, Pennsylvania State University, Princeton University, Stanford University, Yale University, Florida AandM University, Texas AandM University at Corpus Christi Doctoral.

Contacts: *http://www.hfp.heinz.org*

TROPICAL BIOLOGY ASSOCIATION

Offers training courses are intended for students at advanced undergraduate or early postgraduate level who have a keen interest, but little experience, in tropical biology. Applicants must be nationals of an African country. They should have a first degree in a biological science (for example, zoology, botany, and wildlife management or conservation biology) by the time the course begins. Applicants should either be studying for a higher degree or have an active involvement in biological field research; however, applicants should note that the courses are intended for scientists at the start of their research careers. Applications (including letters of reference) must be in English.

Contacts: *The Tropical Biology Association PO Box 44486 – 00100 Nairobi, Kenya*
Phone: 254 20 746090, 749957; Fax: 254 20 741049
http://www.tropical-biology.org

U.S. INSTITUTE OF PEACE

Makes both solicited and unsolicited grants. The Institute's charter is research, education and training, and the dissemination of information on international peace and conflict resolution. The Unsolicited Program is open to any project that fits within the Institute's general mandate of international conflict resolution. Most Institute grants are one to two years in duration.

Contacts: *United States Institute of Peace Grant Program, 1200 17th Street, NW Suite 200, Washington, DC 20036-3011*
Phone: (202) 429-3842; Fax: (202) 429-6063
http://www.usip.org

UNEP-WCMC: CHEVENING SCHOLARSHIPS IN BIODIVERSITY

Training course in Biodiversity Management; One year (no formal degrees awarded). Six scholarships per annum each. In addition one economy class airfare to and from the UK is paid and a small sum available for local travel. Applicants are free to propose biodiversity-related subjects of particular interest to their region, and successful candidates will be encouraged to bring supporting information resources. Candidates must be under 35 years of age, be fluent in English (IELTS band 7 or above), hold a good degree, and preferably also a postgraduate degree, in a biological or environmental science. The Centre does not award formal qualifications and the scheme is not primarily intended for those seeking higher degrees.

Contacts: *The Head of Human Resources, UNEP-WCMC; 219 Huntingdon Road; Cambridge CB3 0DL, UK*
Phone: +44 1223 277314; Fax +44 1223 277136
http://www.unep-wcmc.org

UN-FAO: Visiting Academic and Research Programme

FAO's Visiting Experts Programme, based on institutional arrangements with academic and scientific organizations, provides a framework for distinguished scholars and researchers who contribute their energies to issues related to hunger and food security. The Programme aims to tap the top-level expertise from academic and research institutions of member nations of the Food and Agriculture Organization (FAO) and make them partners in the implementation of the priority programmes of FAO. Highly qualified academics and researchers undertake work in FAO for a maximum of a year on FAO priority subjects, with a view to enhancing the role of FAO as a centre of excellence.

Contacts: *Partnership Programmes; Room B306; Food and Agriculture Organization of the United Nations; Viale delle Terme di Caracalla, 00100 Rome, Italy*
http://www.fao.org

W.E. Upjohn Institute – Grants

Through its grant program, the Upjohn Institute supports policy-relevant research on issues related to employment and unemployment. Grants made under this program are expected to result in research of a rigorous nature that is accessible and of interest to practitioners and policymakers.

Contacts: *W.E. Upjohn Institute for Employment Research; 300 S. Westnedge Avenue; Kalamazoo, Michigan 49007-4686*
Phone: 616/343-5541; Fax: 616/343- 3308
http://www.upjohninst.org

ECOLOGICAL, ENVIRONMENTAL AND CONSERVATION RESEARCH

AFRICA/ASIA LINCOLN PARK ZOO FIELD CONSERVATION FUNDS

The emphasis of the funds is to support new conservation initiatives focused on projects that have: direct impact on wildlife conservation and/or conservation biology; direct participation by graduate students, undergraduate students, or wildlife managers (e.g., wardens, rangers) that enhance their educational experience and improve their opportunities for careers in wildlife management, conservation biology; involvement by students and/or local field assistants from Latin America, Africa, or Asia at levels that engender appreciation for wildlife conservation; links to either the Lincoln Park Zoo animal collection or conservation activities of the zoo staff.

Contacts: *Lincoln Park Zoo Field Conservation Funds, Department of Conservation and Science Lincoln Park Zoo; 2001 North Clark Street; Chicago, IL 60614, USA conservation@lpzoo.org; http://www.lpzoo.com*

AFRICAN BIRD CLUB: CONSERVATION AWARD; EXPEDITION AWARD

The ABC's Conservation Programme supports small conservation projects in Africa. Conservation awards are given to projects surveying and researching African birds, educational projects or training courses, production of guides to the common birds of a country in local languages, interpretation material for nature reserves, and other ideas will be considered. Expedition awards are provided for larger projects. Applicants must be normally resident in Africa; ideally the requested grant should be a major part of the project budget. Requests for contributions towards academic fees are not considered. However funding for field projects that are being carried out as part of an academic study may be considered if funding from the academic institution is not available. The recipient is required to provide a written report on the project results which may be published in the Bulletin of the African Bird Club. Proposals accepted throughout the year.

Contacts: *African Bird Club, c/o BirdLife International, Wellbrook Court, Girton Road, Cambridge CB3 0NA, United Kingdom; http://www.africanbirdclub.org/*

African Fellowship Programme – The Earthwatch Institute

The programme is designed to address the following key training needs in sub-Saharan Africa, identified by Convention on Biological Diversity, and regional studies such as the PARCS report: limited access to modern developments in conservation science and ecology; poor links between conservation professionals from different African countries and lack of understanding between management and research staff.

Contacts: *Earthwatch Europe, 267 Banbury Road, Oxford, OX2 7HT, United Kingdom*
info@earthwatch.org.uk
http://www.earthwatch.org

Alcoa Foundation

The Foundations aims to help communities build partnerships and change lives by actively investing in the quality of life in Alcoa communities worldwide. This involves combining funding with hands-on support to volunteers. The Foundation's grants address global and local needs by partnering with Alcoa communities. Key areas of focus are conservation and sustainability; global education and workplace skills; business and community partnerships; and safe and healthy children and families. Check the website to review the eligibility criteria.

Contacts: *Alcoa Corporate Center*
201 Isabella Street, Pittsburgh, PA 15212-5858 USA
Phone:412-553-4545; Fax: (412) 553-4498
http://www.alcoafoundationfellows.org

Alexander F. Skutch Award

Research grant funding to support the study of life histories, especially social relations and reproduction, of little known birds on the continental neotropics, including Trinidad and Tobago.

Contacts: *http://www.afonet.org/english/skutchapp_english.pdf*

America Philosophical Society

General Research Grants: younger research scholars with at least 1 year of postdoctoral work or the equivalent. Slater Fellowship in the History of Twentieth Century Physical Science. Check website for detailed information on submission deadlines and eligibility requirements.

Contacts: *Independence mall E. 104 S. 5th Street; Philadelphia, PA 19106-3387*
Phone: (215)440-3429
http://www.amphilsoc.org

AMERICAN BIRD CONSERVANCY: WILLIAM BELTON SMALL GRANTS FUND

ABC's William Belton Small Grants Fund will provide support for projects to conserve Endangered and Critically Endangered bird species in Latin America and the Caribbean, in partnership with the U.S. Fish and Wildlife Service. Conservation projects at sites identified by the Alliance for Zero Extinction as essential for the conservation of Endangered and Critically Endangered bird species will be given highest priority for support. ABC Small Grants also provide funding for development of Conservation Plans.

Contacts: *http://www.abcbirds.org/international/small_grants*

AMERICAN LIBRARY ASSOCIATION – WHITNEY CARNEGIE AWARDS

The Whitney-Carnegie Awards, available through the American Library Association, are given for the preparation of guides to research resources in the literature of a chosen discipline or subject area. The award may be used to cover any cost involved in the preparation of such aids, including the cost of research and compilation. Awards do not cover the cost of equipment or production costs. Scholars, librarians, and bibliographers may submit proposals on any generally recognized discipline or subject area. Applications may be submitted year round.

Contacts: *ecotton@ala.org*

AMERICAN MUSEUM OF NATURAL HISTORY

The American Museum of Natural History offers competitive grants and fellowships in areas broadly related to its scientific and educational objectives. These areas include the fields of vertebrate zoology, invertebrate zoology, paleozoology, anthropology, astrophysics and earth and planetary sciences. Monetary awards and deadlines vary according to the particular scholarships, fellowships or grants.

Contacts: *http://www.research.amnh.org/grants/index.html*

AMERICAN ORCHID SOCIETY

The American Orchid Society (AOS) solicits applications from graduate students working on orchid-related thesis projects for the Furniss Foundation/American Orchid Society Fellowship. The purpose of the fellowship is to encourage doctoral candidates in their pursuit of fundamental and applied research on orchids; to attract future generations of scientists to orchidology; and to address critical research needs in the biology of orchids. Fellows spend up to three years working on orchid-related dissertations projects that lead to the Ph.D. degree from accredited institutions. The candidate's dissertation project must deal with any aspect of orchid biology in the disciplines of physiology, molecular biology, structure, systematics, cytology, ecology, or evolution. First refusal rights for all publications arising from work funded by the AOS will be granted to the AOS. Ph.D. candidates may apply. Women, minorities, and persons with disabilities are especially encouraged to apply.

Contacts: *American Orchid Society 16700 AOS Lane Delray Beach FL 33446-4351, United States*
Phone: +1 (561) 404-2050\+1 (561) 404-2000;
Fax+1 (561) 404-2100\+1 (561) 404-2045
pgiust@aos.org

American Ornithologists Union

The American Ornithologist's Union offers research funds for all areas of avian biology for student members of the American Ornithologists Union or members without access to funds from major granting agencies. The committee award about 28-30 grants in of awards up to $1,800. Successful applicants may receive only one research award per year, and only two in a lifetime. Typically the two awards would be for an M.S. project, and, later, for a Ph.D. project. Refer to website for application procedures.

Contacts: *http://www.aou.org*

American Society of Mammologists: Latin American Fellowship

The Latin American Fellowship has been established to promote the support of field research by Latin Americans in Latin America. Eligible students must be citizens of Latin American countries (excluding Puerto Rico), and enrolled in a graduate program in either a Latin American or North American university. Proposed projects must be primarily field oriented with a research emphasis in the areas of natural history, conservation, ecology, systematics, wildlife biology, biogeography, or behavior. The areas of research in mammalogy shall be considered equally important by the selection committee.

Contacts: *http://www.mammalsociety.org/committees/indexlatin.asp*

American Society of Primatologists

Awards and small grants for primate research with awards ranging from $750 – 1,500.

Contacts: *http://www.asp.org/conservation*

Andrew W. Mellon Foundation: Conservation and the Environment

The Andrew W. Mellon Foundation supports grantees within six defined program areas: Higher Education and Scholarship, Scholarly Communications, Research in Information

Technology, Museums and Art Conservation, Performing Arts, Conservation and the Environment. In the Conservation and Environment area, the following grants are accepting proposals: Junior Faculty Research grants are awarded to new faculty as they begin their first tenure-track positions. The awards do not replace start-up research funds provided by universities and are only made after those arrangements are in place. The idea is to provide intellectual venture capital to promising researchers at the critical and formative stage at the beginning of their independent research careers. This program is devoted to basic research on how natural ecosystems work, emphasizing support of leading institutions, innovative research, and the training of promising doctoral and post-doctoral researchers. Grant inquiries should be made, by writing or e-mail, to the appropriate program officer and program area.

Contacts: *The Andrew W. Mellon Foundation; 140 E 62nd St.; New York, NY 10021, USA,*
Phone: (212) 838-8400; Fax: (212) 500-2302
http://www.mellon.org/grant_programs/programs

Animal Behavior Society

The Animal Behavior Society Conservation Committee has compiled a list of more than 60 institutions (with their corresponding www links) that would consider funding research and conservation proposals from behavior-conservation biologists. This is a great source of information for a broad audience, including professors in academia and teaching institutions, graduate and undergraduate students, as well as professionals in conservation-related organizations.

Contacts: *http://www.animalbehavior.org/ABS/Conservation/ccfunding.html*

Animal Welfare Trust

AWT operates on three levels in terms of helping to make a difference to the welfare of animals for Research Grant. First, grant programme that is particularly focused on helping grassroots efforts that have a compelling vision as to how they can make a unique contribution to the animal welfare cause. Second, establishing our own projects, either alone or in partnership with other organizations, that serves the mission to advance the cause of animal welfare. Third, researching, writing and publishing on subjects where animal welfare is compromised and abusive practices can be exposed.

Contacts: *Animal Welfare Trust, Post Office Box 737, Mamaroneck, NY 10543 USA*
Phone:++1(914) 944-3446,
email@www.animalwelfaretrust.org
http://foundationcenter.org/grantmaker/awt/

Annenberg/CPB Project – Funding for New Course Materials

The Annenberg/CPB Project has, since 1981, funded the development of video series that air on PBS and cable and are used in the following ways: (i) by colleges, universities, and high schools as classroom teaching resources; (ii) by colleges and universities as complete telecourses for distant learners; (iii) by PBS viewers and other informal learners for personal enrichment. Those wishing to submit an idea for new course materials may submit a two-page description of the project addressing the criteria below. Submission requirements: 2-page description of project (see website). Submission can be done any time.

Contacts: *The Annenberg/CPB Project; 901 E Street NW Washington, DC. 901 E Street, NW, Washington, DC 20004-2037*
Phone: (202) 879-9600
http://www.cpb.org

Asahi Glass Foundation

The Asahi Foundation supports scientific and technological advances for future generations. Through its research assistance program, it promotes creative research that answers the international community's needs and contributes to achieving breakthroughs in the major issues facing mankind. It funds research in four areas: (i) natural sciences (environment, energy, life sciences, substances and materials, information science); (ii) human and social sciences (environment, organizations, information, humanity); (iii) Comprehensive Research (international or interdisciplinary research on the global environment) and (iv) overseas research assistance. Applicants must be researchers at Japanese universities or affiliated research institutes that offer a graduate-level study programmes. Co-researchers of any nationality are eligible with applications written in Japanese.

Contacts: The Asahi Glass Foundation; 2nd Floor, Science, Plaza 5-3; Yonbancho, Chiyoda-ku, Tokyo 102-0081, Japan
Phone: +81 3 5275 0620, Fax: +81 3 5275 0871
post@af-info.or.jp;
http://www.af-info.or.jp

Ashoka Trust for Research in Ecology and the Environment

ATREE seeks to advance the protection of the environment, conserve biodiversity and promote sustainable use of resources. ATREE administers a small grants programme in Eastern Himalayan region (and to a limited extent also for the Western Ghats region) to contribute in biodiversity conservation needs of the region. ATREE supports activities in the following areas: sustainable natural resources use; capacity building of local institutions; strengthening of the policy framework, environmental education, emergent action/activism, conservation based enterprises and survey and information collection to strengthen biodiversity conservation.

Contacts: *Bungalow No. 2, Bhujiapani Near Airforce Officers' Enclave, 734422, Bagdogra, West Bengal – India,*
Phone: +91-353-550093 or +91-353-551110;
atree@dte.vsnl.net.in;
http://www.atree.org

Association of Zoos and Aquariums

The Association of Zoos and Aquariums' (AZA's) Conservation Endowment Fund (CEF) provides financial support for the conservation and related scientific and educational initiatives of AZA member institutions and their collaborators. Established in 1984, the CEF is a critical part of the AZA's effort to meet its conservation objectives. All principal investigators (PIs) must be individual AZA members. Institutional membership does not count as individual membership. Other participants actively involved in the design or completion of the project should be listed as collaborators and do not have to be AZA members.

Contacts: *AZA Conservation Endowment Fund (CEF)*
Association of Zoos and Aquariums
8403 Colesville Road, Suite 710 Silver Spring MD 20910-3314 United States
cef@aza.org

ATSE Crawford Fund

The Fund promotes and supports international research and development activities in which Australian research organizations and companies are active participants. It supports the work of the Australian Centre for International Agricultural Research (ACIAR), the Australian Agency for International Development (AusAID), and the Consultative Group on International Agricultural Research (CGIAR). The ATSE Crawford Fund Derek Tribe Award was inaugurated in 2001 to promote international agricultural research. The award is made biennially to a citizen of a developing country in recognition of their distinguished contributions to the application of research in agriculture or natural resource management in a developing country or countries.

Contacts: *Senior Administrator Central Office*
Phone: (03) 9347 8427; International: 61 39347 8427
http://www.crawfordfund.org

BAT CONSERVATION INTERNATIONAL STUDENT SCHOLARSHIP PROGRAMME

Each year, BCI sponsors students in conducting conservation-relevant research. Lack of knowledge about bat ecology and behavior is one of the greatest impediments to bat conservation progress. The goal of this program is to support exceptionally talented students in research initiatives that will contribute new knowledge essential to conserving bats and the ecosystems they serve. To this end, BCI has set aside an annually student scholarship programme.

Contacts: *http://www.batcon.org*

BÄUME FÜR MENSCHEN

Bäume für Menschen, which means Trees for the People, is a small foundation (with limited means). Its purpose is the reforestation of endangered or totally deforested areas by means of small, local nurseries. The long-term aim is to ensure the survival of the dry savanna forests. Its activities concentrate on the creation of forest plantations; prevention of further uncontrolled forest exploitation; improvement of agro-forestry systems as well as fruit cultivation within the plantations; education of foresters and rangers; and teaching the local population about the sense and purpose of reforestation.

Contacts: *Angerkapellenstr. 5 (am Landratsamt Nordseite), 82363 Weilheim in OB – Germany,*
Phone:+49-881-8001, Fax: +49-881-8111,
trees@t-online.de
http://www.baeume.de

BAXTER ALLEGIANCE – HEALTH SERVICES PRIZE FOR RESEARCH

Nominations for this prize must provide complete documentation including: a letter of nomination which outlines the nominee's contributions to the health of the public in the three primary focus areas of health services management, health policy development and health care delivery health services; a current curriculum vitae; and two letters of support are required.

Contacts: *HSR Prize Committee, AUPHA, 730 11th Street, NW, 4th Floor, Washington, DC 20001 USA;*
http://www.aupha.org/baxter.htm

BEARDSLEY ZOO

The Beardsley Zoo provides funding to projects involving endangered/threatened species. Funding priorities are given to studies research programmes with significant conservation impacts and education.

Contacts: *Connecticut's Beardsley Zoo, 1875 Noble Avenue, Bridgeport, CT 06610*
Phone: (203) 394-6565
http://www.beardsleyzoo.org

BERGEN COUNTY ZOO

The Bergen County Zoo provides funding to an average of 2.5 projects per year. Funding requirements are conservation impact, time frame and in-situ component with priorities given to endangered/threatened and education in North and South America.

Contacts: *http://www.co.bergen.nj.us*

BERMUDA AQUARIUM MUSEUM AND ZOO

The Bermuda Aquarium Museum and Zoo provides funding to an average of two projects a year with 80% of projects funded in-situ and 20% ex-situ. Funding priorities are endangered/threatened species, in collection, conservation impact, education and exhibit development.

Contacts: *Bermuda Zoological Society P.O. Box FL 145 Flatts FLBX Bermuda;*
Phone: (441)-293-2727; Fax: (441)-293-4014;http://www.bamz.org

BINDER PARK ZOO

The Binder Park Zoo provides funding to an average of one project per year. Funding requirements are endangered/threatened or conservation. Funding priorities are given to in collection and education.

Contacts: *Binder Park Zoo - 7400 Division Drive - Battle Creek,*
MI 49014 (269) 979-1351;http://www.binderparkzoo.org

BIOTECHNOLOGY AND BIOLOGICAL SCIENCES RESEARCH COUNCIL

The Biotechnology and Biological Sciences Research Council (BBSRC) is the UK's principal funder of basic and strategic biological research. To deliver its mission, BBSRC supports research and research training in universities and research centres throughout the UK, including BBSRC -sponsored institutes; and promotes knowledge transfer from research to applications in business, industry and policy, and public engagement in the biosciences. BBSRC's science programmes are administered by seven committees, each dealing with a major area of the Council's scientific remit. Check the website for funding opportunities.

Contacts: *Polaris House; North Star Ave.; Swindon SN2 1UH UK*
Phone: +44 (0)1793 413200; Fax: +44 (0)1793 413201
http://www.bbsrc.ac.uk/funding/Welcome.html

BIRMINGHAM ZOO

The Birmingham Zoo provides funding to an average of 7.5 projects per year with 25% of projects funded in-situ the other 75% are funded ex-situ. Funding priorities are given to conservation impact, holistic, wild cattle and Southeast Asia.

Contacts: *Birmingham Zoo, Inc. 2630 Cahaba Road Birmingham, AL 35223*
Phone: 205-879-0409; Fax: 205-879-9426
http://www.birminghamzoo.com

BRITISH ECOLOGICAL SOCIETY

The following grants are currently offered by BES: *(i) Early Career Project Grants:* The aim of this programme is to assist promising young ecologists by supporting innovative or important research of a pure or applied nature. Initiating or developing research programmes is becoming increasing difficult for ecologists during the early stages of their career, and this programme principally aims to provide an opportunity for ecologists recently appointed to academic posts to establish themselves. *(ii) British Ecological Society (BES) Overseas* Bursaries: scheme provides grants for ecologists in developing countries for innovative ecological research. It aims to encourage links between ecologists in developing countries with those in the UK and other European countries. *(iii) Early Career Project Grants:* The aim of this programme is to assist promising young ecologists by supporting innovative or important research of a pure or applied nature. Initiating or developing research programmes is becoming increasing difficult for ecologists during the early stages of their career, and this programme principally aims to provide an opportunity for ecologists recently appointed to academic posts to establish themselves. *(iv) The Expedition Grants:* Are intended to further the education of young ecologists by extending their experience overseas through assisting student expeditions. No restriction is made on the country(ies) visited, but travel must normally be to or from the United Kingdom or Eire (sixth form expeditions may be within the UK). Expeditions must consist of at least three people (preferably undergraduates or sixth-form students), although participation of postgraduate students and staff is not excluded. It is understood that teaching staff may help in the planning of the expedition, and participate in the trip, but it should be clear that the application is not for a field or adventure training course. The expedition should be distinct from timetabled teaching activities and should primarily have been planned and organized by students. Funding will not be given for work that contributes directly to the higher degree of any of the expedition members. *(v) The Building Capacity for Ecology Fund*: Assists in the establishment or development of networks of ecologists that will allow ecologists in countries that currently lack a well-developed society or ecology institute to interact and gain from the experience and activities of said colleagues. Ultimately, the interactions promoted by the Fund will lead to the creation of new national or regional societies that will affiliate to International Association of Ecology (Intecol).

Contacts: *Grants Officer: dominic@britishecologicalsociety.org*
http://www.britishecologicalsociety.org

British Ornithologist's Union

The British Ornithologists' Union will promote understanding and conservation of the world's birds, advance ornithology within the scientific community and promote scientific ornithology to the wider bird watching public. The BOU has a long tradition of providing financial support for research and expeditions. It administers the David Lack and Landsborough Thomson Trusts, funded by bequests and donations, from which research grants are awarded annually to sponsor scientific and conservation projects. Most of BOU's support is in the form of small grants, but the BOU does from time to time sponsor much larger projects.

Contacts: *http://www.bou.org*

Budweiser Renewable Energy and Wildlife Conservation Research Prize

This new research prize is awarded under a competitive process and recognizes and support a single innovative project that advances new technologies or practices that contribute to making renewable energy a practical energy alternative. The prize is awarded in consideration of a project's ability to develop or evaluate new cost-effective renewable energy technologies for industrial application while demonstrating the measurable benefits such practices can have in advancing wildlife conservation.

Contacts: *Budweiser Renewable Energy Prize; National Fish and Wildlife Foundation; 1120 Connecticut Ave., NW Suite 900; Washington, DC 20036; USA*
RenewableEnergyPrize@nfwf.org
http://www.nfwf.org

Buffalo Zoological Gardens

The Buffalo Zoological Gardens provides funding to an average of 1.5 projects per year. 100% of projects are funded ex-situ. Funding requirements are in collection and time frame. Funding priorities are given to projects involving or containing endangered/threatened loss using well defined protocol and testable hypothesis.

Contacts: *http://www.buffalozoo.org*

Cape Biotech Trust

The Cape Biotech Trust is a Biotechnology Innovation Centre (BIC) funded by the South African Department of Science and Technology with the objective to promote and develop a biotechnology industry in the Western Cape through project investments and capacity development. The Trust provides project investment and capacity development focused on five key areas: diagnostics, vaccines, drug delivery, bio-prospecting and nutraceuticals. This involves identifying and targeting for investment those research and development, businesses, technology platforms and cross cutting infrastructural "building blocks" core to an effective functioning innovation value chain, and building the bio-economy in the Western Cape on the concept of an industry cluster that integrates across the national bio-industry value chains. CBT recommends that all the prospective bio-entrepreneurs have a discussion about their proposals with the CBT project team. This increases the chances of submitting a proposal that is in line with CBT requirements. The submitted proposals are screened for proof of concept and strategic fit with respect to the CBT focus areas. Once it fulfils the stage 1 criteria it will proceed to Stage 2. The Cape Biotech MBA thesis sponsorship programme has also been initiated as a means to encourage and support the development of the cross-over science business skills that are required to support the biotech industry in South Africa.

Contacts: *Cape Biotech Trust; 1st Floor, Minolta House; Black River Business Park; Fir Road, Observatory, 7925 Cape Town, South Africa*
Phone: +27 21 442 3780; Fax: +27 21 442 3781
info@capebiotech.co.za; http://www.capebiotech.co.za

Center for Tropical Forest Science

The Center for Tropical Forest Science is a programme within the Smithsonian Tropical Research Institute that coordinates a pan-tropical network of large-scale Forest Dynamics Plots, each using a standardized protocol. Anyone working directly in a Forest Dynamics Plot, analyzing data from a plot, identifying plants or animals in a plot, or generating complementary data that strengthens Forest Dynamics Plot programs is eligible to apply for a research grant. Projects can be field-oriented, herbarium- or laboratory-based, or analytical. Research projects can be either basic or applied in nature. Social scientists as well as natural scientists are encouraged to apply. Submissions are accepted yearly. The CTFS Grant Program is open to all researchers, from graduate students to senior scientists. Preference will be given to scientists in the countries with CTFS sites and to all graduate students and postdoctoral researchers. Applicants are welcome from all nationalities.

Contacts: *Smithsonian Tropical Research Institute; CTFS Grants Program;*
Unit 0948 APO AA 34002, USA
http://www.ctfs.si.edu

CGIAR CAPRI Grants

CAPRi launched a CGIAR Research Grants Program, providing funding to support innovative empirical research on property rights and collective action issues. The new component addresses a growing demand among members of the CGIAR and their research partners to better understand the nature of institutions of property rights and collective action and how these institutions shape natural resource management and poverty alleviation. Whereas only researchers based at CGIAR Centers may apply for the grants, the program requires partnership with national research institutions and strongly encourages collaboration with other CGIAR centers.

Contacts: *http://www.capri.cgiar.org/resgrants.asp*

CGIAR Challenge Program on Water and Food: IFS Applicants from the Volta Basin

In a joint initiative with the CGIAR Challenge Program on Water and Food, the IFS calls for researchers interested in working in teams on topics related to linkages between water management, food production, poverty, and the environment. This recruitment supports ongoing research activities related to the use and management of small reservoirs and small reservoir ensembles. Successful applicants have the opportunity to work in partnership with others from different research backgrounds and with senior researchers from many regional and international research institutions. Applicants must be from one of the 6 countries of the Volta River Basin; preferably under the age of 40; and not currently associated with a CGIAR research center. Female applicants are especially welcome.

Contacts: *http://www.waterandfood.org*

CGIAR: Challenge Programme Grants

The Consultative Group on International Agricultural Research (CGIAR) has a series of new programs as part of a far-reaching effort to enhance the effectiveness and impact of its research investment. The "Challenge Programs," or CPs, have the following characteristics: (i) address complex issues of great global and/or regional significance; (ii) involve high impact research relating to CGIAR goals; (iii) work through broad partnerships involving a wide range of institutions; (iv) independently governed; (v) committed to achieving specific objectives within an agreed time frame. Four CPs are being implemented. Initial indications from pilot programs showed that the CPs are fulfilling their purpose. The CGIAR has therefore decided to embark on a second cycle with the aim of advancing the new CGIAR system priorities. Programs are chosen for funding via a process of concept note submission, pre-proposals, and full proposals. Check website for updates and future opportunities. Challenge Programs involve at least two CGIAR Centers and two developing country national agricultural research systems (NARS).

Contacts: *CGIAR Secretariat, The World Bank; MSN G6-601; 1818 H Street NW; Washington, DC 20433 USA*
Phone: (202) 473-8951; Fax: (202) 473-8110
cgiar@cgiar.org
http://www.cgiar.org

Chevron Conservation Awards

This is North America's oldest private conservation award program, and it is judged annually by a panel of independent conservationists. The program recognizes outstanding contributions to the conservation of natural resources and offers volunteers, conservation professionals and non-profits organizations alike. Originally created by the late writer Ed Zern in 1954, the program has recognized more than 1,000 conservationists. Recent award winners have been engaged in diverse areas of conservation – from ensuring the survival of wild seahorses, to protecting the world's largest Orangutan population in Indonesia, to providing solar power for health clinics and schools in Africa. See Award website for call for nominations.

Contacts:
http://www.chevron.com/social_responsibility/community/programs_conservation.asp

Cheyenne Mountain Zoo

The Cheyenne Mountain Zoo provides funding of under $1,000 to an average of 1.5 projects per year. 75% of projects are funded in-situ, 25% are funded ex-situ. Funding requirements include endangered/threatened species and habitats, projects with significant conservation impacts and endorsed. Funding priorities are given to projects that include in collection, time frame, in-situ component, black-footed ferret, declining amphibians, felids, okapi and Central and South America.

Contacts: *Delia Garell. 719-633-9925 ext. 120, dgarell@cmzoo.org*

Chevrontexaco Conservation Awards

Program annually recognizes outstanding contributions of individuals and organizations to the conservation of natural resources. Six awards are given per year.

Contacts: *http://www.chevrontexaco.com*

Chicago Zoological Society

The Chicago Zoological Society provides funding of $1,000 to $5,000 to an average of 37.5 projects per year. 90% of funding is provided in-situ. The project must be endorsed to be considered. Funding priorities include endangered/threatened species and habitats, collections management, studies with conservation impacts etc. Applications for grants are by invitation only and are screened carefully to identify projects of the highest priority to the zoo.

Contacts: *Manager, Library Services Chicago Zoological Society*
Brookfield Zoo, 3300 Golf Road, Brookfield, IL 60513
http://www.brookfieldzoo.org

Cincinnati Zoo and Biological Garden

The Cincinnati Zoo and Biological Garden provides funding of $1,000 to $5,000 to an average of 19 projects per year. 94% of projects are funded in-situ. Funding priorities include endangered/threatened species.

Contacts: *http://www.cincinnatizoo.org*

CLEVELAND METROPARKS ZOO

Cleveland Metroparks Zoo provides funding of $1,000 to $5,000 to and average of 50 projects per year. 75% of projects are funded in-situ. Funding priorities include endangered/threatened species, conservation impact, educational, endorsed, time frame, holistic, in-situ component, husbandry, involvement by staff and neotropics. The Scott Neotropical Fund was initiated to provide funds to deserving students and scientists living and working in Latin America. Annual awards are made to research and conservation initiatives involving animals and their habitat or programmes involving local people that directly impact animal conservation. The primary beneficiary of the funds as well as the impact of the project must be within the neotropics (Mexico, Central America, South America and the Caribbean). The Scott Neotropical Fund supports the work of neotropical residents in their countries through direct project support, training opportunities, and/or technical assistance that will continue to benefit the local people, wildlife and habitats of the neotropics into the future.

Contacts: *http://www.clemetzoo.com/conservation/*

CLIMATE CHANGE ADAPTATION IN AFRICA

Canada's International Development Research Centre (IDRC) is a public corporation created by Parliament to help developing countries find practical, long-term solutions to problems they face through applied research. IDRC supports research to help people cope with the effects of climate variability for many years through its Environment and Natural Resources Management program. The Climate Change Adaptation in Africa (CCAA) is a joint program of the International Development Research Centre (IDRC), Canada, and the Department for International Development (DFID), U.K. The CCAA research and capacity development program aims to improve the capacity of African countries to adapt to climate change in ways that benefit the most vulnerable. Building on existing initiatives and past experience, the CCAA programme works to establish a self-sustained skilled body of expertise in Africa that will enhance the ability of African countries to adapt. Supported research in this first round addresses key areas of climate vulnerability in The CCAA program is run and executed in Africa by Africans and, to the extent possible, involving the whole of Africa.

Contacts: *ccaa@idrc.ca*
http://www.idrc.ca

COLUMBUS ZOO AND AQUARIUM

Columbus Zoo and Aquarium provides funding of $1,000 to $5,000 to and average of 80 projects per year. 95% of projects are funded in-situ. Funding priorities include endangered/threatened species, conservation impact, in collection, educational, endorsed, time frame, holistic, in-situ component and benefits to indigenous people. The Conservation Fund serves as a small grants programme that field researchers throughout the world can access by submitting applications directly to the Field Conservation Coordinator. The conservation committee reviews proposals each month.

Contacts: *http://www.colszoo.org/Conservation/*

Company of Biologists Limited

The Company of Biologists will consider direct applications from graduate research students and postdoctoral researchers in the fields covered by the company's journals for travel grants towards the cost of attendance at research conferences or workshops or for skill-acquiring visits to other research laboratories. Applications are particularly welcome from researchers in the field of the company's new journal 'Disease Models and Mechanisms', covering disease models and mechanisms, andin other fields where they can be justified. Applications will not be considered from those eligible to apply for travel grants from the British Society for Cell Biology (BSCB), British Society for Developmental Biology (BSDB), Society for Experimental Biology (SEB), or any other CoB grant-aided source. Applicants must be a postgraduate or junior (in most cases first three years post Ph.D.) postdoctoral researcher. Only one person per laboratory may apply for an overseas visit. For laboratory visits, a letter of support should be included from the host institution. The level of funding is normally £250 for UK and £500 for overseas. The date of travel must occur within at least six months of the grant being made available.

Contacts: *Company of Biologists Limited, Bidder Building 140 Cowley Road, Cambridge CB4 0DL United Kingdom*
http://www.biologists.com/direct_grants.html

Conservation International

CI is a large conservation organization that focuses its work on trying to preserve and promote awareness about the world's most endangered biodiversity through scientific c programs, local awareness campaigns, and economic initiatives. CI prioritizes the protection of biodiversity in tropical forest areas and has on-going programs in 25 countries throughout the world. It assists in the design and implementation of conservation policies and funding programs with multilateral institutions and national governments, and has a number of grant programmes including: *(i) Critically endangered neo-tropical species fund* CENSF aims to contribute to global biodiversity conservation by providing strategically targeted, catalytic support for the conservation of Critically Endangered species living in their natural habitats in the Neo-tropics (here defined as all of Latin America, from Mexico southwards, and including the Caribbean and other offshore islands). Program does not accept unsolicited proposals. *(ii) Margot Marsh Biodiversity Foundation:* Provides strategically targeted, catalytic support for the conservation of endangered nonhuman primates and their natural habitats. Projects submitted to the foundation should have one or more of the following characteristics: focus on critically endangered and endangered nonhuman primates living in their natural habitats. *(iii) Primate Action Fund:* Focuses on critically endangered and endangered nonhuman primates living in their natural habitats; location in areas of high overall biodiversity and under great threat (e.g., "threatened hotspots", "mega-diversity" countries) - to ensure maximum multiplier effect for each project; direction and management by nationals from the tropical countries, to help increase local capacity for implementing biodiversity conservation; the ability to strengthen international networks of field-based primate specialists and enhance their capacity to be successful conservationists; and projects that result in publication of information on endangered primate species in a format that is useful both to experts and the general public. Program does not accept unsolicited proposals. *(iv) Turtle Conservation Fund:* Awards are granted to organizations or individuals for specific conservation or research projects dealing with tortoises or freshwater turtles, but not marine turtles, with either partial or full support as funding allows. Priority for funding is usually given to projects that focus on species that are already highly threatened according to the IUCN Red List criteria, either Critically Endangered or Endangered, or which may be highly threatened but not yet recognized as such. Award recipients enter into contractual agreement with Conservation International to produce the proposed work (CI serves as TCF's umbrella organization). Award recipients are also encouraged to publish at least partial results of the supported research in the international.

Contacts: *2501 M Street, NW Suite 200, DC 20037 Washington DC – USA*
Phone: +1-202-4295660; Fax: +1-202-8870193
http://www.conservation.org

Conservation and Research Foundation

CRF gives priority to unique conservation efforts which are not funded by more conventional funding agencies. Issues supported include the establishment of reserves, development of open space programs, environmental law, preservation of unique habitats, insect control, and the conduct of ecological surveys leading to better land use patterns. Other issues supported include overpopulation, human reproduction. The Foundation also provides funding for individual research projects on biological sciences. Scholarships are not provided. Inquiry regarding the possibility of support should be in the form of an exploratory letter.

Contacts: *Box 5261, Connecticut College, CT 06320 New London– USA.*
http://www.biodiversityeconomics.org/funding/dir3-13.htm

Conservation Technology Support Programme

The ESRI Conservation Program is the non-profit t support arm of the Environmental Systems Research Institute (ESRI). It has helped to create and develop spatial analysis, computer mapping and geographic information systems (GIS) capability among thousands of non-profit organizations and individual projects of all sizes and types worldwide by donating and providing millions of dollars worth of computer technology and training for groups just beginning to work on geographic problems, on an ongoing basis, and for advanced groups at the cutting edge of conservation biology and spatial sciences. Eligible applicants are US-based 501c (3) nonprofit organizations that actively engage the public in resource conservation and environmental protection. This includes, but is not limited to, grassroots conservation and environmental organizations, community action groups, economic development organizations, sustainable development groups, community-based conservation groups, growth management organizations, environmental justice groups, and tribal conservation organizations. International groups must have a US-based tax-exempt organization as their sponsor. CTSP is particularly interested in proposals that address three sectors of sustainability: environment, community and economy.

Contacts: *ESRI Conservation Program; 380 New York St.; Redlands, CA, 92373 USA*
Phone: 909-793-2853 x2488; Fax: 909-793-5953
ecp@esri.com
http://www.conservationgis.org

CONSERVATION, FOOD AND HEALTH FOUNDATION

The purpose of the Conservation, Food, and Health Foundation is to assist in the conservation of natural resources, the production and distribution of food, and the improvement and promotion of health in the developing world. Through grants to support research and through targeted grants to help solve specific problems, the Foundation helps build capacity within developing countries in its three areas of interest: conservation, food and health. The Foundation is especially interested in supporting projects which lead to the transfer of responsibility to the citizens of developing countries for managing and solving their own problems and developing the capacity of local organizations. Preference will be given to projects, including research projects, in areas that tend to be under-funded. Preference will be given to organizations located in developing countries or to developed country organizations whose activities are of direct and immediate benefit t to developing countries.

Contacts: *Conservation, Food, and Health Foundation; c/o Prentice Zinn, Administrator; Grants Management Associates; 77 Summer St, Suite 800; Boston, MA 02110-1006 USA; Phone: 617-426-7080 ext. 307; Fax: 617-426-7087*
pzinn@grantsmanagement.com
http://www.grantsmanagement.com

CONSERVE AFRICA FOUNDATION

CAF was founded in 1998 by a group of NGOs, public institutions and individuals in Africa and in the North, involved in the issues pertaining to sustainable development and conservation in Africa. Its purpose is to work to promote a healthy, safe, fair and sustainable future for all the people of Africa, improving the quality of their life through the advancement of environmental protection and conservation. Through its EcoFund a wide range of community-based initiatives, that protect the environment in the most undeserved and threatened regions in Africa, are supported; including activities to combat desertification.

Contacts: *24 Carterhatch Road, Enfield, Middlesex, EN3 5LS London - United Kingdom; Phone: +44-20-8443.4852; info@conserveafrica.org*
caf@conserveafrica.org; http://www.conserveafrica.org

DALLAS WORLD AQUARIUM

The Dallas World Aquarium provides funding of over $10,000 to an average of three projects per year. 100% of projects are funded in-situ. Funding priorities include endangered/threatened species, conservation impact, educational, time frame, and in-situ component and husbandry.

Contacts: *http://www.dwazoo.com*

DARWIN INITIATIVE

The main objective of the Darwin Initiative is to assist countries rich in biodiversity but poor in resources with the conservation of biological diversity and implementation of the Biodiversity Convention drawing from the British expertise. Projects funded under the Initiative are collaborative, involving either local institutions or communities in the host country. Projects should have a real impact on the ability of the host country to meet its obligations under the Biodiversity Convention. Whenever possible, Darwin funding can be used as a catalyst to lever additional funding for project work, which would not otherwise be forthcoming. The outputs and outcomes from projects should be additional to that from work being funded through other mainstream environmental or research programmes. Projects funded under the Initiative should demonstrate good value for money.

Contacts: *Darwin Applications Management Unit, c/o ECTF; Pentlands, Science Park, Bush Loan; Penicuik, Edinburgh, EH26 0PH UK*
Phone: +44 (0)131 440 5181, Fax: +44 (0)131 440 5501
darwin-applications@ectf-ed.org.uk; http://www.darwin.gov.uk

DAVID AND LUCILLE PACKARD FOUNDATION

Guided by the business philosophy and values of our Founders, the Foundation invests in and take smart risks with innovative people and organizations to improve the lives of children, enable the creative pursuit of science, advance reproductive health, and conserve and restore earth's natural systems. The Foundation focuses the majority of grant-making on three program areas: *(i)* the conservation and science programme that seeks to protect and restore our oceans, coasts, and atmosphere and to enable the creative pursuit of scientific research toward this goal; *(ii)* the population programme that seeks to slow the rate of growth of the world's population, to expand reproductive health options among the world's poor, and to support reproductive rights; and *(iii)* the children, families, and communities programme that seeks to ensure opportunities for all children to reach their potential. Application instructions for each subprogram that accepts letters of inquiry or proposals are available in detailed descriptions of the core programs. Grants are only for charitable, educational, or scientific purposes, primarily.

Contacts: *The David and Lucile Packard Foundation; 300 Second St., Los Altos, CA, 94022 USA*
Phone: (650) 948-7658; http://www.packard.org

DENVER ZOO

The Denver Zoo provides funding of $1,000 to $5,000 to an average of 11 projects per year. 82% of projects are funded in-situ. Funding requirements are conservation impact and holistic. Funding priorities include endangered/threatened species, educational, time frame, in-situ component, keystone/indicator species, habitat conservation, dry and grassland ecosystems, local projects, Mexico and Mongolia.

Contacts: *zooconservation@denverzoo.org*
http://www.denverzoo.org

DETROIT ZOOLOGICAL INSTITUTE

The Detroit Zoological Institute provides funding of over $10,000 to an average of 20 projects per year. 50% of projects are funded in-situ. Funding priorities include endangered/threatened species, conservation impact, educational, endorsed, time frame, holistic, in-situ component and husbandry.

Contacts: *http://www.detroitzoo.org*

DISNEY FUND: WILDLIFE CONSERVATION

The Fund was established in 1995 as a global awards programme for the study and protection of the world's wildlife and ecosystems. It provides annual awards to US non-profit conservation organizations working alongside their peers in other countries. Many of the recipient organizations concentrate their activities on "biological hotspots" - areas rich in plant and animal life at risk of imminent destruction. Since its inception, the Fund has supported more than projects in 20 countries. Each project is evaluated on specific criteria, including scientific methodologies, magnitude of need, involvement of partner organizations, ability to impact an area in the near-term and elements of public education and community involvement.

Contacts: *Conservation Initiatives, Walt Disney World, P.O. Box 10000, Lake Buena Vista, FL 32830 USA*
Phone: (407) 828-3814; Fax: (407) 828-2251
http://disney.go.com/disneyhand/environmentality/dwcf/index.html

DOE - BIOLOGICAL AND ENVIRONMENTAL RESEARCH PROGRAMME

Over the years, the Biological and Environmental Research (BER) Programme has been investing to advance environmental and biomedical knowledge connected to energy. The BER programme provides fundamental science to underpin the business thrusts of the Department's strategic plan. Through its support of peer-reviewed research at national laboratories, universities, and private institutions, the programme develops the knowledge needed to identify, understand, and anticipate the long-term health and environmental consequences of energy production, development, and use. Consistent with the overall objectives, the program conducts numerous research studies to assess directions in science and to identify and assess new and improved approaches to science management.

Contacts: *Human Genome Program: Ethical, Legal and Social Implications*
Director, Grants and Contracts Division; Office of Science, SC-64 U.S
Department of Energy; 19901 Germantown Road; Germantown, MD 20874-1290
http://www.er.doe.gov/production

DURAS PROJECT

DURAS is a GFAR-Agropolis International project supported by the French Ministry of Foreign Affairs through its Priority Solidarity Fund (PSF). It aims to contribute to strengthening the involvement and enhancing the scientific potential of stakeholders in agricultural research for sustainable development (ARSD). Application for the Competitive Grants is open to all stakeholders involved in agricultural research, provided that the Lead Proponent (LP) – a university, national research center or a civil society organization (CSO) - is from a PSZ country covered by the project. This project covers countries from Africa, Near East and some parts of Asia that are listed in France's Priority Solidarity Zone (PSZ).

Contacts: http:*//www.duras-project.net*

EARHART FOUNDATION

Areas of interest for Earhart Foundation include economics, history, international affairs, and political science. Grants are for individual researchers, and each project should have a projected outcome such as teaching, lecturing, or publication. .Submissions can be made anytime

Contacts: *2200 Green Road, Suite H. An Arbor, Michigan 48105*
Phone: 734-761-8592
http://www.dupontfund.org

EARTHWATCH CENTER FOR FIELD RESEARCH GRANTS

Earthwatch works with its scientists to identify appropriate donors, such as companies, institutions, trusts, foundations and government departments, to lever extra funding for additional activities that strengthen the value and integrity of the research. It also enables the project to disseminate the results more widely and build the capacity of organizations, communities and individuals in the research region. Earthwatch preferentially fund projects that specifically focus on one or more of four priority research areas: *(i)* sustainable resource management; *(ii)* climate change; *(iii)* oceans; *(iv)* sustainable cultures. The mission of Earthwatch is to promote sustainable conservation of natural resources and cultural heritage by creating partnerships between scientists, educators, and the general public. Faculty in need of funds and volunteers to support field research may apply to the Center for Field Research (CFR). The Center, in collaboration with its affiliate, EARTHWATCH, supports both basic and applied field research in the sciences and humanities that directly addresses primary sources - whether natural, art factual, or archival. The Center encourages proposals that are interdisciplinary and/or transnational. CFR considers proposals for field research in any discipline that can gainfully employ non-specialists in the implementation of a carefully constructed pure or applied research project. The volunteers, who are recruited and screened to meet scientists' needs, are highly educated citizens dedicated to improving environmental understanding. Specific requests for proposals are available.

Contacts: *For research grant information, contact: Research and Education Department; Earthwatch Institute (Europe); 267 Banbury Rd.; Oxford OX2 7HT UK*
Phone: +44 (0) 1865 318824; Fax: +44 (0) 1865 311383
re@earthwatch.org.uk
http://www.earthwatch.or

The Center for Field Research; 3 Clocktower Place; Suite 100, Box 75; Maynard, Massachusetts 01754; Phone: (978) 461-0081; Fax: (978) 461 2332
cfr@earthwatch.org

EcoLogic Development Fund

The fund was established in 1992 with the objective to conserve endangered wildlife and wild lands by advancing community-based development and resource management. It funds projects that ensure the sound stewardship of biologically rich tropical forests, train farmers in sustainable agriculture, and strengthen community groups so that they can better protect their threatened natural resources. EcoLogic's assistance ranges from helping communities to declare areas protected in order to thwart cattle ranchers and developers, strategic planning to finance reserves and protected areas, providing training in park management, and sponsoring workshops for community leaders on forest co-management. In 1999, EcoLogic Development Fund launched a "green" loan fund that offers fair credit to support environmental enterprise development.

Contacts: *P.O. Box 383405, MA 02238-3405 Cambridge – USA*
Phone: +1-617-441.63.00; Fax: +1-617-441.63.07
enews@ecologic.org; http://www.ecologic.org

Environmental Leadership Programme

The program seeks to transform public understanding of environmental issues by training and supporting a diverse network of visionary, action-oriented emerging leaders. The fellowship programme annually selects approximately twenty individuals with demonstrated talent and potential to participate in a two-year leadership development fellowship. The ELP Fellowship combines a strong peer network and intensive skills training with opportunities for mentoring, project seed money, time for personal and professional reflection, and technical support. ELP Fellows have access to an internal Activity Fund that allows them to pursue capacity building projects to develop leadership skills and accomplish tangible results. Applicants must have approximately three to ten years of work experience or post-undergraduate academic studies related to environmental issues and are residents of the United States or U.S. territories or Freely Associated States. U.S. citizenship is not required.

Contacts: *http://www.elpnet.org/fellowship.html*

Environmental Research and Education Foundation: Francois Fiessinger Scholarship

Awarded in memory of Francois Fiessinger, Ph.D., these annual awards support the work of outstanding doctoral students interested in excellence in environmental research and science. This scholarship is for research within or outside of the U.S. in biological, physical, and social sciences as they relate to environmental research. The scholarship ranges up to $12,000 per year paid monthly, renewable for two additional years for a total of $36,000. Progress must be certified annually by the student's academic advisor. Amounts awarded take into account the cost of tuition at the recipient's institution and any other funds received. Full scholarships cannot be awarded to students who will be receiving full-tuition scholarships from other sources (excluding direct university assistance such as tuition remission and/or income received from assistantships), although such students may still be considered for partial awards.

Contacts: *http://www.erefdn.org/scholar.html*

Environmental Research and Education Foundation

The foundation awards several grants each year for research or education in topics pertaining with any aspect of solid waste management. Project topics could deal with any aspect of the following, but are not limited to:

i. waste generation rates and composition;
ii. waste minimization;
iii. collection and transport;
iv. sorting, recycling, and remanufacture;
v. disposal options (e.g. land filling or incineration);
vi. waste or energy recovery (e.g., composting, landfill gas to energy);
vii. Innovations in collection and transportation equipment development;
viii. Employee health and safety;
ix. Sustainability of resources;
x. Life-cycle assessment of waste management;
xi. Educating corporate customers in purchasing environmentally preferable waste services;
xii. Development of high school and college educational programs.

Grant applications are accepted three times each year, coinciding with the meetings of the Board of Directors.

Contacts: Environmental Research and Education Foundation; 901 N. Pitt Street,
Suite 270; Alexandria, VA 22314 USA
Phone: 703-299-5139; Fax: 703-299-5145
foundation@erefdn.org; http://www.erefdn.org/index.html

EPA PROGRAMMES

(i) *Research opportunities fellowships for graduate environmental study:* The U.S. Environmental Protection Agency (EPA), as part of its Greater Research Opportunities (GRO) programme, offers Graduate Fellowships for masters and doctoral level students in environmentally related fields of study. Master's level students may receive support for a maximum of two years. Doctoral students may be supported for a maximum of three years with funding available, under certain circumstances, over a period of four years. The fellowship program provides up to $37,000 per year of support. *(i) National network for environment management studies:* The purpose of the NNEMS program is to provide students with practical research opportunities and experiences in an EPA office or laboratory, increase public awareness of and involvement in environmental issues, encourage qualified individuals to pursue environmental careers, and help defray the costs associated with the pursuit of academic programs related to the field of environmental protection (e.g. pollution control). The projects are specifically narrow in scope, allowing students to complete the research project while working full-time at EPA during the summer or part-time during the school year. All applicants must be citizens of the United States, its territories or possessions, or lawfully admitted to the United States for permanent residency. These individuals must be pursuing an educational program directly related to pollution control or environmental protection for the duration of the fellowship and currently enrolled in a graduate or Ph.D. program or can provide proof of acceptance and enrollment in a graduate or Ph.D. program at the time of the fellowship award. Applicants must have completed of at least one semester of graduate or Ph.D. work, or at least four undergraduate courses related to the field of environmental studies to be eligible for the fellowship. Recipients of NNEMS fellowships receive a stipend based on their level of education and the duration and location of the project.

(ii) *Science to achieve results (star) for graduate environmental study:* The purpose of the programme is to encourage promising students to obtain advanced degrees and pursue careers in environmentally related fields. Applicants must attend a fully accredited U.S. college or university, and must also be citizens of the US or its territories or possessions, or be lawfully admitted to the US for permanent residence. Students must attend a fully accredited U.S. college or university. Students must be pursuing a master's or doctoral degree in an environmentally related field of specialization. Students who have completed more than one year in the master's programme or four years in the doctoral programmes are not eligible.

(iii) *Tribal lands environmental scholarship:* Full-time Junior, Senior, or Graduate student with a cumulative minimum 2.5 GPA who majors in an environmentally related discipline: biology, biochemistry, chemical engineering, chemistry, entomology, environmental science, hydrology and related disciplines. Applicants must be at least 1/4 American Indian/Alaskan Native and/or recognized member of a federally recognized tribe.

Contacts: *USEPA Headquarters; Office of Research and Development National Center for Environmental Research; Office of the Director; 1200 Pennsylvania Avenue NW; Washington, DC 20460*
http://www. es.epa.gov

ERIE ZOO

The Erie Zoo provides funding of under $1,000 to and average of six projects per year. 75% of projects are funded in-situ. Funding priorities include endangered/threatened species, in collection, conservation impact, endorsed and educational.

Contacts: *eriezoo@erie.net.*
http://www.eriezoo.org

ERIK HOSKING TRUST

Awards are granted annually to applicants pursuing project on birds and other natural history subjects that are of scientific and conservation value, using the media of writing, photography, painting or illustration.

Contacts: *http://www.hosking-tours.co.uk/EricHosking/Bursary.pdf*

EXPLORER'S CLUB

The Explorer's Club offers the Exploration Fund, which provides grants in support of exploration and field research. Grants of up to $1,200 are made primarily to graduate students. Applicants do not have to be members of The Explorers Club and do not have to reside in the United States to qualify for an award. Applications will be judged on the scientific and practical merit of the proposal, the competence of the investigator, and the appropriateness of the budget. Refer to website for application procedures. The Explorer's Club also offers the Youth Activity Fund, which provides grants to full-time high school and undergraduate college students to enable them to participate in field research in the natural sciences under the supervision of a qualified scientist and/or institution. Refer to website for application procedures.

Contacts: *Exploration Fund; The Explorers Club; 46 East 70th St.; New York, NY 10021 USA;*
Phone: 212 628 8383 ext.20; Fax. 212-288-4449
http://www.explorers.org/index.php

ExxonMobil/ESSO: Save the Tiger Fund

The Save The Tiger Fund supports organizations and governmental agencies to conduct a broad range of tiger conservation activities.

Contacts: *http://www.nfwf.org/programs/stf.cfm*

Fannie Mae Foundation

Fannie Mae Foundation's areas of interest include housing policy, housing finance, and community development. Grants are awarded to individuals for empirical or theoretical research. Research is published in its journal, the Housing Policy Debate.

Contacts: *N. Tower, Suite 1 4000 Wisconsin avenue, NW Washington, DC 20016-2800*
Phone: 202-274-8000; Fax: 202-274-8111
http://www.fanniemaefoundation.org

Fauna and Flora International

FFI provides large financial support to conservation initiatives throughout the world, in the form of partnerships, technical assistance, direct funding and consultancy. Funding priority is for protected areas management to support new or existing national parks and reserves, land purchase in areas of high biodiversity in imminent danger of destruction, and for small-scale conservation projects. Additionally, support is offered to ex-situ conservation by captive breeding and plant propagation initiatives and ecological restoration through species re-introduction and invasive species. The Flagship Species Fund of FFI in particular aims to provide practical support to the conservation of endangered species and their associated ecosystems in developing countries.

Contacts: *Great Eastern House, Tenison Road, CB1 2TT, Cambridge - United Kingdom*
Phone: +44-1223-571000; Fax: +44-1223-461481
info@fauna-fl ora.org
http://www.fauna-flora.org

Federal Ministry: Funding for Agricultural Research Centers

The main objective of the Federal Ministry for Economic Cooperation and Development (BMZ) in funding International Agricultural Research is to provide a strategic instrument for enhancing research on scientific and technological innovations concerning sustainable agricultural development, focusing on the impact research has on the poor populations in developing countries. Research support addresses policy issues, the social, environmental and technological dimensions of research, the training of professionals and the dissemination of information.

Contacts: *Federal Ministry for Economic Cooperation and Development*

(BMZ), Division 314, Rural Development, Global Food Security,
Postfach 120322 53045 Bonn – Germany.
Phone: +49 6196 79 1416; Fax: +49 6196 79 7137
http://www.gtz.de/de/dokumente/en-beaf-bmz-guidelines-projectfunding

FOLSOM'S CHILDREN'S ZOO

The Folsom Children's Zoo provides funding of under $1,000 to an average of six projects per year. 50% of projects are funded in-situ. The project must be in collection to be considered. Funding priorities include endangered/threatened species, conservation impact, educational, endorsed, holistic, single species and husbandry.

Contacts: *http://www.lincolnzoo.org*

FORT WAYNE CHILDREN'S ZOO

The Fort Wayne Children's Zoo provides funding of $1,000 to $5,000 to an average of two projects per year. 100% of projects are funded in-situ. Funding requirements are conservation impact, educational, holistic and in-situ component. Funding priorities include endangered/threatened species.

Contacts: *http://www.kidszoo.com*

FOSSIL RIM WILDLIFE CENTER

The Fossil Rim Wildlife Center provides funding of $1,000 to $5,000 to an average of one project per year. 100% of projects are funded in-situ. Funding priorities include endangered/threatened species, conservation impact, educational, time frame, holistic, in-situ component and native Texas wildlife.

Contacts: *http://www.fossilrim.com*

FOUNDATION FOR DEEP ECOLOGY

The Foundation funds activities that focus on *(i)* forests and habitats including campaigns that deal with the root causes of this problems; *(ii)* grassroots activism in areas as diverse as agriculture, economics, bioregionalism; *(iii)* providing aid to traditional native societies to maintain their economic, political, and cultural sovereignty; and *(iv)* campaigns to educate the public about and to resist the growth of new economic trends (e.g. international free trade agreements, etc.). Funding grants vary from small to medium.

Contacts: *Building 1062 Fort Cronkhite, CA 94965, Sausalito – USA*
Phone: +1-415-229.9339; Fax: +1-415.229.9340
info@deepecology.org
http://www.deepecology.org

German Federal Ministry: Small Grants for International Agricultural Research

The Grant programme provides complementing project funding for International Agricultural Research Centers with a simple and rapid instrument to provide comparatively small-scale funding volumes for small project activities. The programme also provides German agricultural research institutions with funding to initiate, prepare, plan, complement or follow up cooperation with International Agricultural Research Centers. The development objectives and criteria applied to this program are identical to those used for 'project funding'. The German 'thematic priorities' are not applied to small grants. Small grants can be requested for innovative projects. In the case of German research institutions, small grants can complement existing collaboration with International Agricultural Research Centers. The project should be around 1 year and not exceed 2 years. An extension is not possible.

Contacts: *http://www.gtz.de*

German Federal Ministry for Environment, Nature Conservation and Nuclear Safety

Germany has put about 60 million Euros into a fund for a "Climate protection initiative" which has a component to support projects which protect biodiversity relevant to protect our climate One particular focus is on the interface between biodiversity conservation and climate protection, for example the conservation of large forests, which is also very important for the climate.

Contacts: *http://www.bmu.de/english*

Gladys Porter Zoo

The Gladys Porter Zoo provides funding of $1,000 to $5,000 to an average of five projects per year. 85% of projects are funded in-situ. Funding priorities include endangered/threatened species, in collection, conservation impact, educational, endorsed and in-situ component. .

Contacts: *http://www.gpz.org*

Global Biodiversity Information Facility: Ebbe Nielsen Prize

This annual Prize was established by the GBIF Governing Board to honor the memory of Ebbe Schmidt Nielsen, who was an inspirational leader in the fields of biosystematics and biodiversity informatics. The Prize is primarily intended to be awarded to a promising researcher in the early stage of his/her career who is combining biosystematics and biodiversity informatics research in an exciting and novel way. This does not preclude researchers at later stages in their career from being nominated where they are able to demonstrate new and novel research in biosystematics/biodiversity informatics. Nomination(s) are submitted by the Head of Delegation of any GBIF Voting or Associate Participant. Individuals or institutions who wish to be nominated or to make a nomination should contact the Head of Delegation of a relevant GBIF Participant or Associate Participant.

Contacts: *GBIF Secretariat; Universitetsparken 15; DK-2100 Copenhagen Ø Denmark; Phone: +45 35 32 14 70; Fax: +4535 32 14 80*
ENPrize@gbif.org
http://www.gbif.org/GBIF_org/prize

Global Conservation Fund

The Global Conservation Fund (GCF) is the first major fund designed to quickly mobilize financial resources to finance the creation, expansion, and long-term management of protected areas in the world's biodiversity hotspots, high-biodiversity wilderness areas, and important marine regions. Resources are allocated towards grant-making for projects that lead to protected area outcomes.

Contacts: *Grant Director, Global Conservation Fund At Conservation International; 1919 M Street, NW, Suite 600; Washington, DC20036 USA*
gcfgrants@conservation.org
http://www.conservation.org

Global Crop Diversity Trust

An increasingly unpredictable and changing climate, and a world population expected to reach 9 billion by 2050, will place unprecedented demands on agriculture. Conserving the vast diversity of crop varieties is the only way to guarantee that farmers and plant breeders will have the raw materials needed to improve and adapt their crops to meet these challenges - and provide food for the future. The Trust's response is to raise a $260m endowment, the interest from which is enough to guarantee the effective conservation – and vitally, the ready availability to those who wish to use it. The endowment ensures that the conservation of this most critical resource is placed forever on a firm foundation. The Trust expects to make the first grants for long-term conservation. These grants will be guided by the regional and crop strategies currently underway, and will be made in accordance with the principles and funding criteria adopted by the Trust.

Contacts: *Global Crop Diversity Trust, c/o FAO; Viale delle Terme di Caracalla; 00153, Rome, Italy*
Phone: +39 06 570 55142; Fax: +39 06 570 54951
info@croptrust.org
http://www.croptrust.org

GOLDMAN ENVIRONMENTAL PRIZE

The Goldman Prize annually honors grassroots environmental heroes from the six inhabited continental regions: Africa, Asia, Europe, Islands and Island Nations, North America, and South and Central America. The Prize recognizes individuals for sustained and significant efforts to protect and enhance the natural environment, often at great personal risk.

Contacts: *http://www.goldmanprize.org*

HARIBON FOUNDATION CONSERVATION RESEARCH GRANTS

Haribon in cooperation with the Critical Ecosystems Partnership Fund awards research grants to fill in gaps in knowledge of threatened species and their habitats, and to build local capacity for research. Proposals for field-based projects are accepted from students, conservationists, enthusiasts and organizations from all over the Philippines. Museum and laboratory-based components may be included. Proposals are evaluated based on: *(i)* relevance to the conservation of threatened species and their habitats; *(ii)* scientific merit and value of the research; *(iii)* capacity of researcher(s) to conduct the study; *(iv)* contribution to the professional development of Filipino researchers.

Contacts: *http://www.haribon.org.ph*

HARRY FRANK GUGGENHEIM FOUNDATION

Supports research that promises to increase understanding of human social problems related to dominance, aggression, and violence. The Foundation places a priority on the study of urgent problems of violence, aggression and dominance in the modern world and also encourages related research projects in neuroscience, genetics, animal behavior, the social sciences, history, criminology, and the humanities which illuminate modern human problems.

Particular questions that interest the Foundation concern violence, aggression, and dominance in relation to social change, the socialization of children, inter-group conflict, drug trafficking and use, family relationships, and investigations of the control of aggression and violence. Grants are made for individual projects, not for institutional programmes. Grants may cover such items as salaries, employee benefits, computer time, supplies and equipment, field expenses, research assistance, clerical assistance and any other expenses directly related to the specific research project proposed

Contacts: *Harry Frank Guggenheim Foundation, 527 Madison Avenue; New York, N.Y. 10022-4304; Phone: (212) 644-4907; Fax: (212) 644-5110; http://www.hfg.org*

HEALTH EFFECTS INSTITUTE

HEI is a public private partnership established to research and provide impartial information about the health effects of emissions from motor vehicles and other sources of environmental pollution. HEI receives half of its core funds from the EPA and half from twenty-eight manufacturers or marketers of motor vehicles or engines in the US. The HEI research programme has addressed many important questions about the health effects associated with exposure to both regulated pollutants such as carbon monoxide, ozone, and nitrogen dioxide, and unregulated pollutants such as diesel exhaust, aldehydes, and methanol. HEI has also funded studies to understand the mechanisms of diseases, to develop better methods to assess health effects and determine dose response relationships. The programme has included theoretical, in vitro, animal, controlled human exposure, and epidemiological studies.

Contacts: *Health Effects Institute, Charlestown Navy Yard; 120 Second Avenue, Boston, MA 02129-4533, USA*
Phone: (617) 886-9330; FAX (617) 886-9335
http://www.healtheffects.org

Heifer Project International

Heifer Project helps people utilize livestock as an integral component of sustainable agriculture and holistic development. Heifer Project works at the grassroots level by providing animals and training to organized local groups that request assistance. Projects strengthen rural families and communities through improved nourishment, increased production and the dissemination of skills and knowledge for self-reliance. Care for the earth's natural resources is emphasized through training in livestock management, pasture improvement, soil conservation, reforestation and water harvesting.

Contacts: *P.O. Box 8058, AR 72203 Little Rock - United States*
Phone: +1-501-907.2600;
info@heifer.org
http://www.heifer.org

Henry J. Kaiser Family Foundation

The Henry J. Kaiser Family Foundation supports programmes designed to improve health in the U.S. The Foundation makes more the $24 million in philanthropic expenditures each year. Grant making concentrates on government's role in health and on the health of low-income and minority groups. Within these broad interests, the main areas in which the Foundation currently focuses the majority of its Grant making are: Health Policy, Media and Public Education, and Health and Development in South Africa. Interested applicants may submit a preliminary Letter of Intent at any time. Formal proposals will be invited.

Contacts: *http://www.kff.org/*

Honolulu Zoo

The Honolulu Zoo provides funding of under $1,000 to an average of four projects per year. 100% of projects are funded in-situ. Funding priorities include endangered/threatened species, in collection, conservation impact, educational, endorsed, single species and in-situ component.

Contacts: *151 Kapahulu Ave, Honolulu, HI 96815*
http://www.honoluluzoo.org

IDEA WILD

IDEA WILD's promoted conservation of biodiversity by providing biologists and educators working in developing countries with equipment and supplies to aid in their conservation efforts. Recipients often share equipment with at least 3 other conservation projects. The average cost per project is around $720. With IDEA WILD, a nominal conservation investment goes a very long way!

Contacts: *http://www.ideawild.org*

IMPUMELELO INNOVATIONS AWARDS FOR SOUTH AFRICA

The Impumelelo Innovations Award Trust rewards exceptional projects involving partnerships with the public sector that enhance the quality of life of poor communities in innovative ways. The aim of the Trust is: to improve the quality of life of the poor; to identify and highlight innovative and effective examples of service delivery in the country; to encourage good governance; to break through fiscal and structural delivery constraints; to highlight models of innovative government projects in order to encourage replication; to recognize public managers who are the backbone of these exemplary programmes; and to provide compelling and credible portraits of the many ways in which government contributes to public problem solving and quality of life.

Contacts: *P O Box 1739, Cape Town, 8000 South Africa*
Phone: +27 21 461-3783; Fax: +27 21 461-1340
info@impumelelo.org.za
http://www.impumelelo.org.za

INAMORI FOUNDATION: THE KYOTO PRIZE

The Kyoto Prize is an international award to honor those who have contributed significantly to the scientific, cultural, and spiritual betterment of mankind. The Prize is presented annually in each of the following three categories: Advanced Technology, Basic Sciences, and Arts and Philosophy. Candidates for the Kyoto Prize are nominated by official Kyoto Prize nominators, who are selected annually by the Inamori Foundation from among recognized domestic and international authorities.

Contacts: *620 Suiginya-cho, Shimogyo-ku, Kyoto 600-8411, Japan*
Phone: 075-353-7272; Fax: 075-353-7270
http://www.inamori-f.or.jp

INDIANAPOLIS ZOO

The Indianapolis Zoo provides funding of over $10,000 to an average of two projects per year. 100% of projects are funded in-situ. Funding priorities include in collection, conservation impact, time frame and single species.

Contacts: *http://www.indianapoliszoo.com*

INTERNATIONAL CENTRE FOR GENETIC ENGINEERING AND BIOTECHNOLOGY

The International Centre for Genetic Engineering and Biotechnology conducts innovative research in life sciences for the benefit of developing countries. It strengthens the research capability of its Members through training and funding programmes and advisory services and represents a comprehensive approach to promoting biotechnology internationally. The Centre is dedicated to advanced research and training in molecular biology and biotechnology and holds out the prospect of advancing knowledge and applying the latest techniques in the fields of: biomedicine; crop improvement; environmental protection/remediation; biopharmaceuticals and bio-pesticide production. Funding opportunities are made available through the Collaborative Research Programme (CRP), which is a unique source of funding aimed at financing those projects addressing original scientific questions that show a potential contribution of particular relevance for the applicant's country. Affiliated Centre status is essential in order for laboratories in ICGEB Member States to be eligible to apply for funding. See website for specific deadlines.

Contacts: *ICGEB; Padriciano 99; 34012 Trieste – Italy*
Phone: +39-040-3757382; Fax: +39-040-3757361
http://www.icgeb.trieste.it

INTERNATIONAL CRANE FOUNDATION

The International Crane Foundation provides funding of $1,000 to $5,000 to an average of ten projects per year. 100% of projects are funded in-situ. Funding requirements are in collection, conservation impact and time frame.

Contacts: *http://www.savingcranes.org/awards.html*

INTERNATIONAL DEVELOPMENT RESEARCH CENTER: RURAL POVERTY AND ENVIRONMENT

Rural Poverty and Environment (RPE) program initiative. RPE supports research focusing on the needs of the rural poor who live in fragile or degraded ecosystems. The program initiative uses an approach that combines participatory action research to generate knowledge; capacity development for researchers and decision makers to participate in multi-stakeholder processes; and policy engagement to build action and learning oriented partnerships.

Contacts: *IDRC; RPE Program Initiative; PO Box 8500; Ottawa, Ontario Canada, K1G 3H9*
Phone: (613) 236-6163, Fax: (613) 567-7749
http://www.idrc.ca

INTERNATIONAL DEVELOPMENT RESEARCH CENTRE

IDRC is a public corporation created by the Canadian government to support research conducted by scientists working in universities, private enterprise, governments, and NGOs. They have several programmes related to natural resource management, water, and land use and funding can vary from medium to large. Their community-based natural resource management programme seeks to provide assistance to people living in ecosystems that face increasing resource exploitation to manage and use their natural resources sustainably. Other priorities include food security, equity in natural resource use, biodiversity conservation, sustainable employment, strategies and policies for healthy societies, information and communication, gender, technology, and the environment. In Africa, special focus is on effective use of information and communication technologies.

Contacts: *PO Box 8500, K1G 3H9, Ottawa – Canada;*
Phone: +1-613-236-6163; Fax: +1-613-238-7230
http://www.idrc.ca

INTERNATIONAL FOUNDATION FOR SCIENCE

Applications for IFS Research Grants are welcome from young scientists in developing countries to do research on the sustainable management, use or conservation of biological or water resources. This broad statement covers natural science and social science research on agriculture, soils, animal production, food science, forestry, agro-forestry, aquatic resources, natural products, water resources, etc. To qualify for IFS funding, research projects must be: related to the sustainable utilization, conservation or management of the biological or water resource base; conducted in a developing country; of a high scientific standard; feasible; relevant for the country/region.

Contacts: *International Foundation for Science (IFS);*
Karlavägen 108, 5th floor; SE- 115 26 Stockholm, Sweden
Phone: +46 8 545 818 00; Fax: +46 8 545 818 01
http://www.ifs.se

INTERNATIONAL FUND FOR AGRICULTURAL RESEARCH

IFAR is a results-oriented foundation that supports partnerships, collaboration, and awards as a means of fostering scientific excellence in agriculture and related fields. It aims to: (i) recognize and promote scientific excellence involving the work of international agricultural research centers supported by the Consultative Group on International Agricultural Research (CGIAR) and their partners; (ii) promote science through professional development; (iii) give priority to Africa and South Asia. IFAR annually provides small grants to professionals from national agricultural research systems in developing countries (NARS). Applicants are encouraged to collaborate with CGIAR Centers' scientists when preparing their project proposals for submission.

Contacts: *http://www.ifar4dev.org*

International Human Dimensions Programme: Seed Grant Initiative

The International Human Dimensions Programme on Global Environmental Change is an international, interdisciplinary science programme dedicated to promoting, catalyzing and coordinating research on the human dimensions of global environmental change. IHDP takes a social science perspective on global change and it works at the interface between science and practice. Grants awarded are committed to the production of National Inventories of human dimensions research in selected countries and toward the organization of national or regional workshops for human dimensions researchers to facilitate the creation of new or strengthening of existing research networks.

Contacts: *International Human Dimensions Program on Global Environmental Change*
Walter-Flex-Str. 3, 53113 Bonn, Germany
Phone: +49-228-73 90 50. Fax: +49-228-73 90 54
http://www.ihdp.uni-bonn.de

International Institute for Environment and Development: Drylands Programme

The programme aims to promote the sustainable, equitable, participatory and decentralized management of natural resources in dryland Africa, through policies and institutions, thereby contributing to improved livelihoods, poverty reduction, and long term ecological and economic sustainability. The programme's work covers: land rights and tenure; supporting pastoral civil society; enhancing dryland livelihoods and market opportunities in a globalizing world. Besides implementing projects, IIED complements the ongoing work of partners in Africa (CSOs, development agencies, research institutions) by disseminating research findings, brokering meetings, facilitating dialogue, organizing training, and lobbying activities as well as a small grants facility. The work focuses on the dryland belt of Africa south of the Sahara, particularly the, several coastal countries in West Africa, East Africa and the Horn.

Contacts: *4 Hanover Street, EH2 2EN Edinburgh – United Kingdom*
Phone: +44-131-226.7040; Fax: +44 131 624.7050
drylands@iied.org
http://www.iied.org

International Nutrition Foundation

The fellowship programme for centers of excellence for research on nutrition and infection in developing countries programme's mandate of the foundation is to strengthen a limited number of stable centers of excellence for research related to the interactions of nutrition and infection. It is designed to assist institutions that already have a strong research record or research potential to improve their competence in any field related to this topic. Individual applications for fellowship will not be accepted.

Contacts: *International Nutrition Foundation; 150 Harrison Avenue, Room 232;*
Boston, MA 02111 USA
Phone: (617) 636-3771; Fax: (617) 636-3727;
http://www.inffoundation.org

International Primatological Society

IPS offers small monetary grants to those conducting projects congruent with the aims of our society. Martha J.Galante -- grant proposals are solicited from professionals of habitat countries of primates. Money awarded could be used for conservation training. Lawrence Jacobsen -- Education Committee of IPS solicits grants of up to $1000 to support the development of primate conservation education programs. These initiatives should support field conservation programs, work with local community and/or schools, or are used to provide training in conservation education techniques. Charles Southwick -- this award is dedicated to recognizing individuals living in primate habitat countries that have made a significant contribution to formal and informal conservation education in their countries. Captive Care -- grants of up to $1000 for projects focusing on captive care issues.

Contacts: *http://www.pin.primate.wisc.edu/ips/*

International Research Fellowship Programme

The objective of the International Research Fellowship Program (IRFP) is to introduce scientists and engineers in the early stages of their careers to research opportunities abroad, thereby furthering NSF's goal of creating a diverse, competitive, and globally-engaged U.S. workforce of scientists, engineers, technologists and well-prepared citizens. These awards are available in any field of science and engineering research and education supported by NSF. Foreign science or engineering centers and other centers of excellence in all geographic regions are eligible host institutions.

Contacts: *http://www.nsf.gov*

International Society of Arbori culture

Since 1975, the TREE Fund through the ISA Research Trust has awarded grants to aid, stimulate and encourage scientific studies of trees. The granting programs include *The Hyland*

R. Johns Grant Program. The goal of the all TREE Fund Grant Programs is to provide support for research and technology transfer projects that are in keeping with the TREE Fund mission and priorities, and address topics that have the potential of benefiting the everyday work of arborists. Research and educational projects funded by the TREE Fund must be in keeping with this mission statement. *The Sponsored Grant Program*: Sponsored grants are funded by a sponsor or donor for a specific project or topic. Proposals for the program are limited to the size specified in the "Request for Proposal" (RFP).

Contacts: *TREE Fund; 711 E. Roosevelt Rd.; Wheaton, IL 60187 USA*
Phone: (630) 221-8127; Fax: (630) 690-0702.
http://www.treefund.org

IREX

The International Research Exchange programme has a number of specific sub-programmes of different interest include The John J. and Nancy Lee Roberts Fellowship Programme that supports cutting-edge research in the social sciences on Eastern Europe, the New Independent States (NIS), the Near East, and Asia.

Contacts: *International Research and Exchanges Board (IREX)*
1616 H Street, NW; 6th Floor; Washington, DC 20006; Street, New York, NY 10017
Phone: (202) 628-8188; Fax: (202) 628-8189
irex@irex.org
http://www.irex.org

ITI Scotland Ltd: ITI Life Sciences

ITI Life Sciences, a division of ITI Scotland Ltd., focuses on driving innovation in the life sciences sector. ITI Life Sciences will consider proposals to commission pre-competitive market-driven research programmes that ultimately stimulate entrepreneurial activity in Scotland. ITI Life Sciences is currently investigating several areas of technology, and welcome interest in these areas. Emerging Drug Targets: ITI are interested in opportunities that may arise due to the changi

Contacts: *http://www.itilifesciences.com*

Jennifer Altman Foundation

The Jennifer Altman Foundation is a private foundation dedicated to the vision of a socially just and ecologically sustainable future through program interests in environmental health and mind-body health. The Foundation's interests include: Environmental Health: Presently, this work is focused primarily, but not exclusively, on supporting the emergence, both nationally and internationally, of an informed citizen movement for environmental health. The strategic focus of the Foundations is primarily on the impact of endocrine disrupting chemicals and other fetal contaminants on human health and on biodiversity. We are concerned with human made chemicals that disrupt the developing fetal endocrine system and/or disrupt the developing fetus (human or nonhuman) by affecting systems other than the endocrine system. Specific projects concern: Health Care Without Harm; The International Treaty Process to Eliminate 12 Highly Toxic Chemicals; The International Environmental Health Network; Biotechnology, Chemicals and Environmental Health.

Contacts: *http://www.jaf.org*

Jessie Smith Noyes Foundation Environment Grants

The Jessie Smith Noyes Foundation is committed to protecting and restoring Earth's natural systems and promoting a sustainable society by strengthening individuals, institutions and communities pledged to pursue those goals.

Contacts: *Jessie Smith Noyes Foundation 6 East 39th Street; 12th Floor; New York, NY 10016*
Phone: (212) 684-6577; Fax :(212) 689-6549
http://www.noyes.org

Jessup and McHenry Awards

Jessup funds are awarded competitively to students wishing to conduct studies at the postgraduate, doctoral and postdoctoral levels under the supervision or sponsorship of a member of the curatorial staff of the Academy. The Awards are not available for undergraduate study. Such studies may be in any specialty in which these curators have expertise.

Contacts: *http://www.acnatsci.org/research/jessupinfo.html*

John Ball Zoological Society

The John Ball Zoological Society provides funding of under $1,000 to an average of four projects per year. 25% of projects are funded in-situ. Funding requirements include endangered/threatened, conservation impact, time frame and husbandry. Funding priorities include educational, endorsed and holistic.

Contacts: *John Ball Zoo, 1300 W. Fulton, Grand Rapids MI 49504*
Phone: 616-336-4301; Fax: 616-336-3907; http://www.johnballzoosociety.org

John G. Shedd Aquarium

Funding requirements are conservation impact and educational. Funding priorities include endangered/threatened species, in collection, endorsed, holistic, single species, in-situ component, husbandry and high profile.

Contacts: http://*www.sheddaquarium.org*

JRS Biodiversity Foundation

The Foundation seeks to enhance knowledge and promote the understanding of biological diversity for the benefit and sustainability of life on earth. Interdisciplinary activities primarily carried out via collaborations in developing countries and economies in transition. The Foundation seeks to advance projects, or parts of biodiversity informatics projects that focus on: (i) collecting data, (ii) aggregating, synthesizing, publishing data, and making it more widely available to potential end users, and (iii) interpreting and gaining insight from data to inform policy-makers.

Contacts: *Harry E. Cerino, Executive Director; J.R.S. Biodiversity Foundation*
138 West Highland Avenue; Philadelphia PA 19118 USA
Phone: + 1-267-286-7840; Fax: +1 215 248 2381;
hcerino@jrsbdf.org
http://www.jrsbdf.org

Keidanren Nature Conservation Fund

Keidanren's membership includes all major Japanese companies and industry associations. It is considered one of the most powerful organizations in Japan. The activities of the Fund are: support of activities, including aid, for foreign and Japanese NGO-administered nature conservation projects in developing countries; providing training for personnel that will work on international nature conservation projects; and promotion of environment consciousness. The KNCF supports field projects which provide assistance for nature conservation efforts in developing countries worldwide but particularly in the Asia-Pacific region. The KNCF gives priority to projects that will serve to enrich biodiversity through regeneration of water sources and forests by the local people efforts, conservation efforts, and sustainable natural resources management.

Contacts: *Kotani-Bldg. 8F, 1-2-7 Uchikanda, Chiyoda-Ku, 100-0047 Tokyo – Japan*
Phone: +81-3-5282-5701; Fax: +81-3-5282-5703
kncf@keidanren.or.jp
http://www.keidanren.or.jp/kncf

Kilby Awards

The mission of the Kilby Awards Foundation is to identify, celebrate and provide heroic role models for future generations, the leaders of the 21st Century. The Kilby Awards Foundation annually recognizes five to seven individuals who make extraordinary contributions to society through science, technology, innovation, invention and education. Laureates are chosen without regard to race, religion, gender or national origin.

Contacts: *The Kilby International Awards Fndn; P.O. Box 9109; Dallas, TX 75209 USA*
kilby@kilby.org
http://www.kilby.org

LAKE VICTORIA RESEARCH INITIATIVE

The Inter University Council for East Africa (IUCEA) is pleased to announce the 5th round of research competition under the Lake Victoria Research (VicRes) Initiative funded by Sida/ SAREC supports scientific research contributing The VicRes grant is awarded to a team of PhD holders or one headed by a PhD holder from the five East African Community (EAC) Partner States for inter- and multi-disciplinary projects. Researchers participating in ongoing VicRes research projects do not qualify for this competition. A research team can apply for funding in any one of the following three themes-(i) wetlands (ii) land use and (iii) water catchments management and conservation. The proposed topics may fall within the following clusters: ethno botany and indigenous knowledge; aquaculture technologies; water pollution/heavy metals; resource management and planning and, sustainable land use options focusing on biological, chemical or physical as well as socio-economic, cultural and political aspects of poverty reduction and environmental restoration. Priority will be given to projects that are gender sensitive and are likely to contribute to wealth creation and technological innovations.

Contacts: *Lake Victoria Research (VicRes) Initiative; co IUCEA; 3rd Floor,*
EADB Building
P.O. Box 7110; Kampala, Uganda
coordinator@vicres.net

LARSON AQUATIC RESEARCH SUPPORT

The Larson Aquatic Research Support scholarship provides support for doctoral and Masters students interested in careers in the fields of corrosion control, treatment and distribution of domestic and industrial water supplies, aquatic chemistry, and/or environmental chemistry.

Contacts: *http://www.awwa.org/About/scholars/index.cfm*

LASPAU: FULBRIGHT-OASECOLOGY INITIATIVE

The program offers grants to individuals from Latin America, the Caribbean, and Canada for masters and doctoral level study at U.S. universities. Grantees in the natural sciences, social sciences, and public policy study multidisciplinary approaches to environmental preservation and sustainable development. The goal of the initiative is to develop a well-prepared cadre of environmental professionals who, upon completion of their studies, will return to their institutions to share their expertise with colleagues while maintaining contact with each other.

Contacts: *http://www.laspau.harvard.edu/grant-LASPAU.htm*

LAWRENCE FOUNDATION

The foundation is focused on making grants to support environmental, education, health, human services and other causes. The followings are eligible: U.S. citizens, permanent residents, or foreign nationals currently working legally in the United States; Inventors who have created a product, process or material; made a technology more affordable, redesigned a system, or otherwise demonstrated remarkable technological inventiveness in addressing local or global sustainability in the United States and/or abroad; Inventors who can provide evidence that their inventions have been adopted for practical use; Inventors working in an area that ultimately improves the quality of air, water or soil; or pertains to health, energy, agriculture, shelter, biodiversity or ecosystem management; Individuals who are inspirational to young people, through their creativity, outreach or mentoring activities.

Contacts: *530 Wilshire Blvd., Suite 207 Santa Monica, CA 90401 USA*
Phone: +1 (310) 451-1567; Fax +1 (310) 451-7580
http://www.thelawrencefoundation.org

LEMELSON-MIT AWARD FOR SUSTAINABILITY

Lemelson-MIT Award for Sustainability is bestowed to inventors whose products or processes are viable and sustainable, and have high potential to improve the quality of life for future generations. The award serves to increase awareness of local or global sustainability issues and inventors working in these critical areas, and also supports continued inventive work of these individuals.

Contacts: *Lemelson-MIT Program Officer; The Lemelson-MIT Program; Massachusetts Institute of Technology, School of Engineering; Massachusetts Ave., Building E60-215; Cambridge, MA 02139-4307 USA*
Phone: 617-253-3352; Fax: 617-258-8276
lemelson_awards@mit.edu
http://www.mit.edu/invent/a-main.html

LEVERHULME TRUST

The Trust makes awards for the support of research and education. The Trust emphasizes individuals and encompasses all subject areas. The Trustees place special weight on: the originality of the projects put to them; the significance of the proposed work; the ability to judge and take appropriate risk in the project; the removal of barriers between traditional disciplines. The Trust's financial support is organized into grants and awards which vary in size, purpose and application procedure.

Contacts: *Leverhulme Trust; 1 Pemberton Row, London, EC4A 3BG UK*
Phone: 020 7822 5220; Fax: 020 7822 5084
http://www.leverhulme.ac.uk

LINDBERGH FOUNDATION

Each year, The Charles A. and Anne Morrow Lindbergh Foundation presents Lindbergh Grants to individuals whose proposed research or education projects make important contributions toward improving the quality of life by balancing technological advancements and the preservation of our environment. Grants are made in numerous areas of special interest to Charles and Anne Lindbergh, including aviation/aerospace, agriculture, arts and humanities, biomedical research and adaptive technology, conservation of natural resources, education, exploration, health and population sciences, intercultural communication, oceanography, waste disposal management, water resource management, and wildlife preservation.

Contacts: *The Charles A. and Anne Morrow Lindbergh Foundation*
2150 Third Avenue North, Suite 310; Anoka, MN 55303-2200
Phone: (763) 576-1596; Fax: (763) 576-1664;
info@lindberghfoundation.org
http://www.lindberghfoundation.org

LIVING DESERT

The Living Desert provides funding of $1,000 to $5,000 to an average of 3.5 projects per year. 100% of projects are funded ex-situ. Funding priorities include endangered/threatened species, in collection, endorsed, in-situ component, North and South Africa and Arabian Peninsula.

Contacts: *http://www.livingdesert.org*

LIZ CLAIBORNE AND ART ORTENBERG FOUNDATION

The Foundation concerns itself particularly with matters of species extinction, habitat destruction and fragmentation, resource depletion and resource waste. It favors solutions that directly benefit local communities and serve as exemplars for saving species and wildlands. It recognizes the imperative to reconcile nature preservation with human needs and aspirations. There are two primary programme interests: mitigation of conflict between the land resource needs of local communities; and conservation of biological diversity in rural landscapes outside of parks and reserves; and the implementation of relevant, field-based scientific, technical, and practical training programs for local people. The hallmarks of a Foundation project include local participation, personal involvement and accountability.

Contacts: *Program Manager, 650 5th Avenue, NY 10019 New York - United States*
Phone: +1-212-333.2536; Fax: +1-212-956.3531
lcaof@fcc.net
http://www.lcaof.org/home.html

LOS ANGELES ZOO

The Los Angeles Zoo provides funding of $1,000 to $5,000 to an average of 10 projects per year. 70% of projects are funded in-situ. Funding priorities include endangered/threatened, in collection, conservation impact, time-frame.

Contacts: *http://www.lazoo.org*

LOWRY PARK ZOO

The Lowry Park Zoo provides funding of over $10,000. 100% of projects are funded in-situ. Funding requirements are endangered/threatened species, in-situ component and holistic. Funding priorities include in collection, conservation impact, educational, endorsed, time frame, tropics and subtropics.

Contacts: *http://www.lowryparkzoo.com*

MACARTHUR FOUNDATION RESEARCH AND WRITING GRANTS COMPETITION

Research and Writing Grants Competition of MacArthur Foundation aims to support projects that explore the development of improved understanding of key topics in global security and sustainability, and to broaden and strengthen the community of writers and scholars engaged in work on these issues.

Contacts: J*ohn D. and Catherine T. MacArthur Foundation*
Office of Grants Management; 140 S. Dearborn Street; Chicago, IL 60603-5285
Phone: (312) 726-8000; Fax: (312) 920-6258
4answers@macfound.org
www.macfound.org

MARINE CONSERVATION BIOLOGY INSTITUTE

In the face of increasing evidence that the world's oceans are in trouble, the Marine Conservation Biology Institute (MCBI) established this grant. The program funds research that documents historic conditions in the oceans prior to large-scale human impacts, thereby providing a context for current ecosystem conditions and helping inform management approaches. MCBI invites individuals and collaborative teams from academic institutions and nongovernmental organizations to apply. Preference will be given to graduate students, postgraduate researchers, and early career scientists.

Contacts: *Marine Conservation Biology Institute 14301 Arnold Dr., Suite 25 Glen Ellen California, 95442, United States*
larissa@mcbi.org
http://www.mcbi.org

Max and Anna Levinson Foundation

Funding is evenly distributed among three categories: The Environment: including Protection of Ecosystems and Biological Diversity; Alternative Energy and Conversion from the Oil Economy; Alternative Agriculture and Local Green Economic Development; Breaking the Link Between Resource Extraction, Civil and International Conflict, and Markets; and the Development of Environmental Movements. Social: the Promotion of a More Democratic, Equitable, Just and Rewarding Society, including World Peace; Protection of Civil and Human Rights; Citizen Responses to Globalization; Alternative Media and Education; Community-Based Economic Development; Youth Leadership; and Violence Prevention and Response. Jewish/Israel: including Jewish Culture, Religion, and Spirituality; Yiddish; Building Jewish Community in the Diaspora; Jewish Organizations for Social Change; and Peace, Social, and Environmental Issues in Israel.

Contacts: *The Max and Anna Levinson Foundation; P.O. Box 6309; Santa Fe, NM 87502 USA*
Phone: (505) 995-8802; Fax: (505) 995-8982
info@levinsonfoundation.org;
http://www.levinsonfoundation.org

Memphis Zoo

The Memphis Zoo provides funding of $5,000 to $10,000 to an average of 12.5 projects per year. The project must include conservation impact. Funding priorities include endangered/threatened species, in collection, educational, endorsed, time frame, holistic, single species, in-situ component, husbandry, and meets or promotes the Memphis Zoo's mission statement.

Contacts: *http://www.memphiszoo.org.*

Miami Metrozoo

The Miami Metrozoo provides funding of $1,000 to $5,000 to an average of 10 projects per year. 40% of projects are funded in-situ. Funding priorities include endangered/threatened species.

Contacts: *http://www.miamimetrozoo.com*

Mill Mountain Zoo

The Mill Mountain Zoo provides funding of under $1,000 to an average of 2.5 projects per year. 60% of projects are funded in-situ. Funding priorities include endangered/threatened species.

Contacts: *http://www.mmzoo.org*

Minnesota Zoo

The Minnesota Zoo provides funding of under $1,000 to an average of 1.5 projects per year. 50% of projects are funded in-situ. Project must include conservation impact. Funding priorities include endangered/threatened species.

Contacts: *http://www.mnzoo.com*

Mistra—The Foundation for Strategic Environmental Research

The Foundation for Strategic Environmental Research, Mistra, supports strategic environmental research with a long-term perspective, aiming to solve major environmental problems. The main part of Mistra's funding is focused on broad-based interdisciplinary programmes. Calls for proposals are issued regularly. Check website under "Fundings" for most recent calls.

Contacts: *http://www.mistra.org*

NASA: Research Opportunities in Space and Earth Sciences

NASA's Ocean Biology and Biogeochemistry program focuses on describing, understanding, and predicting the biological and biogeochemical regimes of the upper ocean as determined by remote observation of aquatic optical properties from space, aircraft, and other suborbital platforms. Participation in this program is open to all categories of U.S. and non-U.S. organizations, including educational institutions, industry, nonprofit institutions, NASA Centers, and other government agencies. Historically Black Colleges and Universities (HBCU's), other minority educational institutions, and small businesses and organizations owned and controlled by socially and economically disadvantaged individuals or women are particularly encouraged to apply. Participation by non-U.S. organizat

Contacts: *http://www.nspires.nasaprs.com*

National Fish and Wildlife Foundation

The Foundation fosters co-operative partnerships to conserve fish, wildlife, plants, and the habitats on which they depend. Its Challenge Grant Program awards funding to projects that address priority actions promoting fish and wildlife conservation.

Contacts: *http://www.nfwf.org*

National Geographic: Research and Exploration Fund

The National Geographic Society awards grants for scientific field research and exploration through its Committee for Research and Exploration. All proposed projects must have both a geographical dimension and relevance to other scientific fields and be of broad scientific interest. The committee is emphasizing multidisciplinary projects that address environmental issues (e.g., loss of biodiversity and habitat, effects of human-population pressures). Funding is not restricted to United States citizens. Researchers planning work in foreign countries should include at least one local collaborator as part of their research teams.

Contacts: *http://www.nationalgeographic.com*

National Institutes of Health: Global Research Initiative Programme

The purpose of this initiative is to provide funding opportunities for the increasing pool of foreign social and behavioral scientists, clinical investigators, nurses and other health professionals, with state-of-the-art knowledge of research methods. Research related to the health effects of human exposures to environmental agents is encouraged.

Contacts: *Division of International Training and Research*
Fogarty International Center; 31 Center Drive, MSC 2220; Building 31, Room B2C39; Bethesda, MD 20892-2220
Phone: (301) 496-4596, Fax: (301) 402-0779; http://www.grants.gov

NAOAA: Coral Reef Conservation Programme

The purpose of the International Coral Reef Conservation Program is to support the National Action Plan to Conserve Coral Reefs which calls on the United States and to exercise global leadership in the international arena in shaping and developing environmentally sound and comprehensive coral reef policy, strengthen international conventions and foster strategic partnerships with other countries, international organizations and institutions, the public and private sectors, and non-governmental organizations to address international threats to coral reef ecosystems.

Contacts: *NOAA Coral Reef Conservation Program*
10102, NOAA National Ocean Service, 1305 East-West Highway, Silver Spring, MD 20910.
http://www.grants.gov

National Research Foundation of South Africa

The objective of the National Research Foundation (NRF) is to support and promote research through funding, human resource development and the provision of the necessary research facilities, in order to facilitate the creation of knowledge, innovation and development in all fields of the natural and social sciences, humanities and technology. In so doing, it contributes to the improvement of the quality of life of all the people of the country.

Contacts: *National Research Foundation; PO Box 2600; Pretoria 0001 South Africa*
Phone: +27 12 481 4000 / 4001
http://www.nrf.ac.za/

National Science Foundation: Advance Programme

The NSF funds ADVANCE, which supports the increased participation of women in the scientific and engineering workforce. Members of underrepresented minority groups and individuals with disabilities are especially encouraged to apply. Proposals that address the participation and advancement of women from underrepresented minority groups are encouraged. Institutional Transformation Grants of the Foundation: grants are used to help build structural framework within an institution that will draw girls and women into science and engineering.

Contacts: *http://www.nsf.gov*

National Science and Technology Forum Awards

These awards are unique in South Africa, recognizing the outstanding contributions of individuals and groups of individuals to SET. It affords an opportunity for recognition and celebration to all practicing scientists, engineers and technologists across the system of innovation, including, for example, teachers and students in mathematics, science and technology.

Contacts. *NSTF Secretariat; P.O. Box 9823; Pretoria, Gauteng 0001 South Africa*
Phone: +27 (012) 841-3987 ext. 2708, Fax: +27 (012841-3025
http://www.nstfawards.org.za

Neotropical Bird Club

Grants are available for conservation work or research which may be of conservation benefit in the neo-tropics.

Contacts: *http://www.neotropicalbirdclub.org*

Netherlands Committee for IUCN

NC-IUCN is an independent foundation which unites Dutch NGO-members of IUCN. It promotes the implementation of IUCN's objectives in and by the Netherlands. The Committee runs programmes that include: *(i) the tropical rainforest programme (TRP)* directed at implementing the tropical rainforest policy of the Dutch government; *(ii) the small grants for wetlands programme (SWP)* funds that supports projects in the field of conservation and sustainable management of wetlands, set up and implemented by non-governmental, local organizations (NGOs); *(iii) NC-IUCN-NL Dry and Sub-Humid areas small grants programme* that aims to alleviate poverty through the conservation, restoration and sustainable management of dry and sub-humid ecosystems. Other aims are capacity building of local NGOs (especially dryland ecosystem management) and improved policy influencing by NGOs on all levels. Regarding the sustainable use and conservation of dry and sub-humid ecosystems and their resources a whole range of initiatives is supported: traditional management and conservation systems; community-based management; sustainable use of (wildlife) resources; gender relations and natural resource management; support for ecosystem restoration (forest landscape restoration, natural regeneration); agro-forestry; access to natural resources (including conflict resolution between different stakeholders); marketing of sustainably harvested dry land products; and support for sustainable income generating activities.

Contacts: *PO Box 18184, 1001 ZB Amsterdam – the Netherlands*
Phone: +31-20-626.17.32; Fax: +31-20-6279349
http://www.nciucn.nl

Netherlands Organization for Scientific Research

(i) *African partnership for capacity development and clinical interventions against poverty-related diseases:* The general objective is to strengthen the African research and development capacity in the field of poverty-related diseases by investing in joint research activities and clinical testing in Africa at locally owned research and development. NACCAP thus combines the regular NWO focus on scientific excellence, competitive funding and involvement of the scientific community with capacity development to fight poverty-related diseases;

(ii) *WOTRO* focuses on scientific research on development issues, in particular poverty alleviation and sustainable development. Its scope includes all scientific disciplines and low and middle-income countries. Acting at the interface of science, policy and practice, WOTRO develops, facilitates and supports activities that strengthen the use and impact of scientific research on development issues. These include activities which contribute to: networking, joint agenda-setting, and knowledge-sharing.

(iii) *The integrated programme (IP)* scheme aims at supporting excellent problem-oriented interdisciplinary research to the benefit of development and societal issues in the South. An IP scheme may address developmental and societal issues in a broad sense that may lead to new insights, theories and/or breakthroughs in current knowledge and understanding or the scheme may fit within one or more of four themes of WOTRO's framework: poverty and hunger; global health and health systems; sustainable environment and global relationships.

Contacts: *P.O. Box 93120; 2509 AC The Hague; The Netherlands*
Phone: +31 (0)70 344 0907, Fax: +31 (0)70 381 9874
http://www.nwo.nl

NIEH

Investigator-initiated Research Project Grants areas of Special Emphases include: basic molecular mechanisms of environmental insult genetic susceptibility and predisposition (environmental genome project); human health effects of complex mixtures; reproductive health; neurodegenerative/neurobehavioral diseases or disorders; translational research; impact of environmental exposures on special populations (women, children and minorities); diet and nutrition; immune system modulation; socioeconomic status and health across life course.

Contacts: *http://www.niehs.nih.gov*

NORTH CAROLINA ZOO

The North Carolina Zoo provides funding of over $10,000 to an average of 5 projects per year. 100% of projects are funded in-situ. Funding priorities include time frame, local endorsement and involvement of local participants in countries where projects are conducted.

Contacts: *http://www.nczoo.org*

NORTH EASTERN WISCONSIN ZOO

The North Eastern Wisconsin Zoo provides funding of $1,000 to $5,000 to an average of two projects per year. 100% of projects are funded ex-situ. Funding priorities include endangered/threatened species, conservation impact, educational, endorsed projects, in-situ component and husbandry.

Contacts: *http://www.newzoo.org*

NATIONAL SCIENCE FOUNDATION

The National Science Foundation promotes and advances scientific progress in the United States by competitively awarding grants and cooperative agreements for research and education in the sciences, mathematics, and engineering. The National Science Foundation funds research and education in most fields of science and engineering. It does this through grants, and cooperative agreements to more than 2,000 colleges, universities, K-12 school systems, businesses, informal science organizations and other research organizations throughout the United States. Check website to investigate possible funding in your field on study.

Contacts: *http://www.nsf.gov*

Oklahoma City Zoo

The Oklahoma City Zoo provides funding of $1,000 to $5,000 to an average of four projects per year. 90% of projects are funded in-situ. Funding priorities include endangered/threatened species, conservation impact, educational, time frame, single species, in-situ component and staff involvement.

Contacts: *http://www.okczoo.com*

Oregon Zoo

The Oregon Zoo provides funding of $1,000 to $5,000 to an average of 12 projects per year. 100% of projects are funded in-situ. Funding requirements are in-situ component and conservation impact. Funding priorities include endangered/threatened species, endorsed with time frame.

Contacts: *http://www.oregonzoo.org*

P3 Award: A Student Design Competition for Sustainability

The P3 Award Competition has two phases: initially teams compete for $10,000 P3 grants. Recipients use the money to research and develop their projects during the academic year. Then in spring, the P3 grant recipients are invited to Washington, D.C. to compete for the P3 Award which conveys additional funding for further design development and implementation.

Contacts: *http://www.es.epa.gov/ncer/rfa/*

Packard Foundation, David and Lucile

Provides grants to non-profit organizations in the following broad program areas: conservation, population, science, children, families, communities, arts, and organizational effectiveness and philanthropy. In its international conservation program, the Foundation supports programs which protect globally outstanding habitats in areas of natural significance and biological diversity; address important factors of environmental degradation found in inappropriate and unsustainable land, water, energy, and marine resources; elevate the role of natural and social science in the management of key resources and in the arena of conservation policy.

Contacts: *300 Second Street, Suite 200, CA 94022, Los Altos – USA*

Phone: +1-650-948.7658

http://www.packfound.org

PADI Foundation

The PADI Foundation encourages and supports underwater science, environmental projects, and education. The Foundation funds and assists worthwhile projects that enrich mankind's understanding of the aquatic environment and encourage sensitivity to and protection of the delicate ecological balance of underwater life. The Foundation also funds worthwhile projects to increase understanding of sport diving physics and physiology that benefit the general diving public and add to the scientific understanding of man's relationship and ability to survive in the underwater environment.

Contacts: *9150 Wilshire Blvd., Suite 300; Beverly Hills, CA 90212 USA*
Fax: (310) 859-1430
http://www.padifoundation.org

Palm Beach Zoo at Dreher Park

The Palm Beach Zoo at Dreher Park provides funding of over $10,000 to an average of five projects per year. 80% of projects are funded in-situ. Funding priorities include endangered/threatened species, in collection, conservation impact, endorsed, tropics and subtropics.

Contacts: *http://www.palmbeachzoo.org*

People's Trust for Endangered Species

People's Trust for Endangered Species makes funds available for wildlife researchers and conservationists to carry out crucial work on globally threatened species throughout the world. Projects may either focus on research into the status of a particular species or involve practical conservation work within the field. Priority is given to those projects that are likely to lead to a lasting benefit to the species concerned. PTES also provides grants for students carrying out practical conservation projects and research in the summer months of their undergraduate degree.

Contacts: *http://www.ptes.org*

Phoenix Zoo

The Phoenix Zoo provides funding of $1,000 to $5,000 to an average of 5.5 projects per year. 80% of projects are funded in-situ. The project must include conservation impact. Funding priorities include time frame, holistic, regional conservation value, watershed or landscape level, Sonoran desert and Mid-Southwest.

Contacts: *http://www.phoenixzoo.org*

Pittsburgh Zoo and PPG Aquarium Conservation Fund

The Pittsburgh Zoo Conservation Fund is dedicated to the preservation of wildlife and wild habitat. Annual awards of $1,000 to $3,000 will be made. A variety of projects will be considered, but field studies and cross-disciplinary approaches to conservation are especially encouraged.

Contacts: *http://www.pittsburghzoo.org*

Primate Conservation, Inc.

PCI is a non-profit foundation, which funds research that supports conservation programmes for wild populations of primates. Priority is given to projects that study the least known and endangered species in their natural habitat. The involvement of citizens from the country in which the primates are found is desirable.

Contacts: *http://www.primate.org/*

Pro Regenwald

Pro Regenwald seeks to protect forests, especially tropical forests, and the indigenous peoples who inhabit them. It provides small to medium funds for projects that seek to reverse tendencies which destroy forests and to develop protection measures through education, lobbying, research, and other initiatives. It also has a special fund for small projects especially as starting aid for initiatives and a fund for travel costs of people from the South who want to attend seminars or conferences in Europe.

Contacts: *Frohschammerstr. 14, 80807, Munich – Germany*
Phone: +49-89-3598650; Fax: +49-89-3596622
http://www.regenwald.org

Professional Engineers in Industry Scholarship

Applicants must be sponsored by a NSPE/PEI member. Students who are children, dependents, or relatives of NSPE members are given preference in the scholarship selection process. Students must have completed a minimum of two semesters or three quarters of undergraduate engineering studies (or be enrolled in graduate study) in a program accredited by the Accreditation Board for Engineering and Technology (ABET).

Contacts: *http://www.nspe.org*

Rainforest Alliance

The Alliance is a non-profit organization dedicated to the conservation of tropical forests. It runs the Catalyst Grant Program which is awarded to carefully selected conservation projects where the infusion of a modest amount of money can dramatically increase the effectiveness of a specific endeavor.

Contacts: *65 Bleecker Street, NY 10012, New York – USA*
Phone: +1-212-677.1900; Fax: +1-212-677-2187
canopy@ra.org
http://www.rainforest-alliance.org

Rainforest Concern

This is a registered charity established in 1993 with the specific objective of protecting seriously threatened habitats in Latin America. RC purchases, leases and manages, for protection, threatened native forest with exceptional biodiversity. If land is purchased this is usually done in the name of partner organizations or local communities. Much of the work concerns the creation of protected corridors of forest between existing reserves. Furthermore RC develops programmes for health, education and alternative income generation with local people. These are designed to reduce the human impact on native forests. A significant part of alternative income is derived from responsible ecotourism. Also research in biodiversity is promoted through research facilities at several of the projects.

Contacts: *27 Lansdowne Crescent, W11 2NS, London - United Kingdom*
Phone: +44-171-2292093; Fax: +44-171-2214094
rainforest@gn.apc.org
http://www.rainforest.org.uk

Rainforest Conservation Fund

This is an all-volunteer non-profit organization that is dedicated to preserve the world's tropical rain forests. RCF funds small projects related to rain forests – that is, in or near rain forests. They are, therefore, amenable to projects elsewhere if they are pertinent to rain forest conservation. RCF supports projects aimed at fauna management, community management of lakes, habitat protection and recuperation, agro-forestry, successful species, enriched fallows, maintain local/community possession. The goal is to avoid cutting primary forest and to restore secondary forest and degraded areas.

Contacts: *2038 North Clark Street, Suite 233, IL 60614-4713, Chicago – USA*
Phone: +1-773- 975.75.17
rcf@interaccess.com
http://www.rainforestconservation.org

Ramsar Small Grants Fund

The Ramsar Small Grants Fund was established by Ramsar COP4 in 1990 as a mechanism to assist developing countries and those with economies in transition in implementing the Convention and to enable the conservation and wise use of wetland resources. Suitable project proposals are those which contribute to the implementation of the Convention's Strategic Plan 2003-2008 for the conservation and wise use of wetlands; provide emergency assistance for Ramsar sites; or provide 'preparatory assistance' to allow non-Contracting Parties to progress toward accession. Successful proposals receive 80% of the allocated funds upon signature of the contract and the remainder upon submission of an adequate final report, but countries from which adequate final project reports have not been received may be denied further consideration for funding until those outstanding project dossiers have been closed.

Contacts: *Small Grants Fund (SGF), Ramsar Convention Secretariat;*
28 Rue Mauverney; H-1196 Gland, Switzerland
Fax: +41 22 999 0169
ramsar@ramsar.org
http://www.ramsar.org

Reid Park Zoo

The Reid Park Zoo provides funding of under $1,000 to an average of two projects per year. 100% of projects are funded in-situ.

Contacts: *http://www.tucsonzoo.org*

Research Fellowship Program - WCS-International

Research Fellowship of WCS is a small grant programme designed to support individual field research which is based on sound and innovative conservation science and aims to build capacity for the next generation of conservationists. Professional conservationists from the country of research, and/or Post-graduates pursuing a higher degree. RFP preferably supports projects for only one year, and preference is given to discrete short-term projects.

Contacts: *Programme Co-ordinator, Research Fellowship Program, International Conservation*
Wildlife Conservation Society, 2300 Southern Blvd. Bronx, NY 10460 U.S.A
http://www.wcs.org

Riverbanks Zoo and Garden

The Riverbanks Zoo and Garden provides funding of under $1,000 to an average of seven projects per year. 50% of projects are funded in-situ. Funding priorities include conservation impact, endorsed, time frame and holistic.

Contacts: *http://www.riverbanks.org*

ROCHE RESEARCH FOUNDATION

Through various types of grants, the Roche Research Foundation (RRF) promotes the international standing of Swiss research by supporting scientific investigations by promising young researchers in the areas of biology, chemistry and medicine. The foundation provides funding for researchers and clinicians based abroad to visit Swiss institutions (universities, equivalent research institutes, and hospitals) in order to collaborate on scientific projects, or the development of new techniques or clinical treatments. Such funding is given in terms of fellowships which are generally limited to a few weeks or months, depending on the purpose of the visit, and which do not exceed one year. RRF offers start-up and bridging assistance to researchers or research groups at Swiss universities, equivalent research institutes, or hospitals through fellowships to doctoral and postdoctoral students or, if sufficient funds are available, provides support for the acquisition of equipment indispensable for the planned study and not part of an institute's standard equipment, as well as special consumables that cannot be covered by other sources. The foundation supports research collaborations consisting of two or more researchers in different disciplines at Swiss universities, equivalent research institutes, or hospitals on interdisciplinary life science projects of particular importance to biology, chemistry or medicine. This funding consists primarily of fellowships to young researchers, doctoral students or postdoctoral students. In special cases, the foundation can also fund the procurement of equipment indispensable for the planned study and not part of the standard equipment of at least one of the institutes involved in the research collaboration, as well as consumables not covered by other sources. The foundation awards fellowships to young researchers, undergraduate, graduate and postdoctoral students who lack sufficient funds to continue their experimental scientific work, provided this work is conducted at Swiss universities, equivalent research institutes, or hospitals, or by Swiss nationals abroad. The foundation funds study visits by researchers working in Switzerland to foreign institutions to acquire new methods or specific techniques that are necessary for their work and inaccessible elsewhere. As a rule, the duration of such visits is limited to a maximum of four weeks. To be eligible, a candidate must be a Swiss national engaged in experimental biomedical research at home or abroad; or be a non-Swiss biomedical researcher with the opportunity to work in Switzerland. Recipients must have distinguished themselves in their fields with original and important contributions or possess specialized methodological knowledge and must currently hold an active position in a laboratory or hospital.

Contacts: *http://www.research-foundation.org*

ROCKEFELLER FOUNDATION AND BROTHERS FUND

The Rockefeller Foundation was established in 1913 by John D. Rockefeller, Sr., to "promote the well-being" of humanity by addressing the root causes of serious problems. The Foundation works around the world to expand opportunities for poor or vulnerable people and to help ensure that globalization's benefits are more widely shared. The Foundation is supporting people, institutions and scientific innovation in sub-Saharan Africa and parts of Asia to dramatically increase food production by generating new crop varieties, developing more dependable and sustainable soil conservation and enrichment techniques, and improving access to markets where small farmers can buy seeds and fertilizer and sell any surplus harvest. The foundation also works with public- and private-sector partners to extend the benefits of new agricultural knowledge to those who need them most. The Rockefeller Brothers Fund supports sustainable development grant-making endeavors encourages government, business, and civil society to work collaboratively on environmental conservation and to make it an integral part of all development planning and activity. Recognizing the global nature of many environmental problems, the Fund also promotes international cooperation in addressing these challenges. Sustainable Development: Through this program, the fund seeks to: (i) support strategies to combat global warming and seeks to contribute to promote international cooperation on this issue; and (ii) conserve terrestrial and marine biodiversity by protecting and restoring ecosystems and by fostering sustainable communities that pursue locally appropriate development plans.

Contacts: *The Rockefeller Foundation; 420 5th Ave.; New York, NY 10018-2702 USA*
Phone: (212) 869-8500
http://www.rockfound.org

Rockefeller Brothers Fund, Inc.; 437 Madison Ave., 37th Fl.
New York, NY 10022-7001 USA
Phone: (212) 812-4200; Fax: (212) 812-4299
info@rbf.org
http://www.rbf.org

Roger Williams Park Zoo

The Roger Williams Park Zoo provides funding of $5000 to $10,000 to an average of seven projects per year. 70% of projects are funded in-situ. Funding priorities include endangered/threatened species. Further information provided on the website.

Contacts: *http://www.rogerwilliamsparkzoo.org*

Rosamond Gifford Zoo at Burnet Park

The Rosamond Gifford Zoo at Burnet Park provides funding of under $1,000 to an average of one project per year. 100% of projects are funded in-situ. Funding priorities include in collection studies.

Contacts: *http://www.rosamondgiffordzoo.org*

Royal Caribbean: Ocean Fund

Royal Caribbean Cruises Ltd, which operates Royal Caribbean International and Celebrity Cruises, established the Ocean Fund in 1996 to support marine conservation organizations in preserving the world's oceans. The mission of the Ocean Fund is to support efforts to restore and maintain a healthy marine environment, minimize the impact of human activity on this environment, and promote awareness of ocean and coastal issues and respect for marine life. Ocean Fund grants are made to a variety of nonprofit groups and institutions conducting activities directly related to marine conservation.

Contacts: *Ocean Fund Manager 6th Floor 1050 Caribbean, Way Miami, FL 33132-2096 USA*
Phone: (305) 539-6574
http://www.royalcaribbean.com

ROYAL GEOGRAPHICAL SOCIETY EXPEDITION RESEARCH GRANTS

The Society's grants are wide-ranging and offer support to individual researchers, expedition teams and schoolteachers. Grants are for work in both the UK and overseas. The Society has many small grants for scientific expeditions working in remote and challenging environments.

Contacts: *http://www.rgs.org/*

ROYAL SOCIETY FOR THE PROTECTION OF BIRDS

Each year, the Royal Society for the Protection of Birds (RSPB), the BirdLife Partner in the UK, offers grants to fund research on 190 bird species listed by IUCN as "Critically Endangered". Only works on these species are considered for grants. Priority is given to people working in-country, particularly BirdLife Partners and other conservation NGOs, although individuals not attached to an institution may also apply if they can obtain a letter of support from a recognized institution in their country.

Contacts: *Paul Donald, paul.donald@rspb.org.uk*
http://www.rspb.org.uk

ROYAL SOCIETY: PFIZER AWARD

This award, sponsored by Pfizer Ltd., is designed to reward scientists working in the UK or Africa at the outset of their career and to promote science capacity building in the developing world. The award will recognize a research scientist, at an early stage of their career that is making an innovative contribution to the biological sciences, including basic medical science, which has led or has the potential to lead to a sustainable positive impact on sub-Saharan Africa. Grants are for the recipient to carry out a program of activities that is linked to an African centre of scientific excellence, which would normally be a University, or equivalent, research centre. The award is intended for researchers at the outset of their career with the potential to build a program of activities to follow on from the prize.

Contacts: *The Royal Society, 6-9 Carlton House Terrace, London SW1Y 5AG UK*
Phone: +44 (0)20 7451 2500, Fax: +44 (0)20 7930 2170
http://www.royalsoc.ac.uk

RUFFORD SMALL GRANTS FOR NATURE CONSERVATION

Rufford Small Grants for Nature Conservation (RSGs) are aimed at small conservation programmes and pilot projects. RSGs are available to individuals and small groups. Projects are about a year's duration. Applications from non-first world countries are strongly encouraged.

Contacts: *http://www.rufford.org/rsg*

RUSSELL SAGE FOUNDATION

The Russell Sage Foundation is the principal American foundation devoted exclusively to research in the social sciences. The Foundation is a research center, a funding source for studies by scholars at other academic and research institutions, and an active member of the nation's social science community. Applications are accepted any time.

Contacts: *Program Officer/contact, 112 East 64th Street New York, New York 10021 http://www.epn.org/sage/rsappl.html*

RUTH L. KIRSCHSTEIN NATIONAL RESEARCH SCIENCE AWARDS

The Congress of the United States enacted the National Research Service Act (NRSA) Program in 1974 to help ensure that a diverse pool of highly trained scientists will be available in adequate numbers and in appropriate research areas to carry out the Nation's biomedical and behavioral research agenda.

In 2002, the National Research Service Award Program was renamed the Ruth L. Kirschstein National Research Service Award Program as a tribute to Dr. Kirschstein's years of exceptional service to the Nation.

The National Institutes of Health (NIH) awards individual postdoctoral fellowships to promising applicants with the potential to become productive, independent investigators in fields related to the mission of the NIH constituent institutes and centers. By the time of award, candidates for the postdoctoral fellowship award must be citizens or non-citizen nationals of the United States, or must have been lawfully admitted to the United States for Permanent Residence.

Contacts: *http://www.grants1.nih.gov*

SAINT LOUIS ZOO

The Saint Louis Zoo provides funding of $5,000 to $10,000 to an average of 11 projects per year. 100% of projects are funded in-situ. The project must include Saint Louis Zoo staff members as a participant or sponsor.

Contacts: *http://www.stlzoo.org*

San Antonio Zoo

The San Antonio Zoo provides funding of $1,000 to $5,000 to an average of four projects per year. 75% of projects are funded in-situ. Funding priorities include educational, holistic and husbandry.

Contacts: *http://www.sazoo-aq.org*

San Francisco Zoo

The San Francisco Zoo provides funding of $5,000 to $10,000 to an average of four projects per year. 100% of projects are funded in-situ. Funding priorities include endangered/threatened species.

Contacts: *http://www.sfzoo.org*

Science and Practice of Ecology and Society Award

The Science and Practice of Ecology and Society Award is an annual award given to the individual or organization that is the most effective in bringing trans-disciplinary science of the interactions of ecology and society into practice.

Contacts: *School of H man Evolution and Social Change*
Arizona State University; Box 872402, Tempe, AZ 85287-2402 USA
http://www.ecologyandsociety.org

SeaWorld and Busch Gardens Conservation Fund

The SeaWorld and Busch Gardens Conservation Fund, a non-profit charitable organization, was created in 2003 in order to increase and strengthen the long-standing commitment of Busch Entertainment Corp. (BEC) to the cause of wildlife conservation. Through the SeaWorld and Busch Gardens Conservation Fund, guests at the Anheuser-Busch Adventure Parks, BEC's corporate partners, and the general public are able to join with BEC to support wildlife conservation around the globe.

Contacts: *SeaWorld and Busch Gardens Conservation Fund; 231 S. Bemiston Ave., Suite 600; Clayton, MO 63105 USA.*
mail@swbgfund.org; http://www.swbg-conservationfund.org

SeaWorld, Inc.

Seaworld, Inc. provides funding of over $10,000 to an average of 15 projects per year. 75% of projects are funded in-situ. Project must include time frame, and priorities include endangered/threatened species, in collection, conservation impact and marine vertebrates.

Contacts: *http://www.seaworld.com*

Sedgwick County Zoo

The Sedgwick County Zoo provides funding of $1,000 to $5,000 to an average of five projects per year. 100% of projects are funded in-situ. Funding priorities include endangered/threatened species, in collection, educational, endorsed, time frame, holistic, single species and in-situ component.

Contacts: *http://www.scz.org*

Seneca Park Zoo

The Seneca Park Zoo provides funding of $1,000 to $5,000 to an average of one project per year. 100% of projects are funded ex-situ. Funding priorities include endangered/threatened species, in collection, time frame, holistic, in-situ component, husbandry and upstate.

Contacts: *http://www.senecaparkzoo.org*

Shared Earth Foundation

The Foundation funds organizations that promote protection and restoration of habitat for the broadest possible biodiversity, which foster respect for other species and individual creatures, which work to limit detrimental human impact on the planet, and which further the inherent right of all creatures to share the Earth. The Foundation looks to fund primarily, though not exclusively, small organizations.

Contacts: *The Shared Earth Foundation; 113 Hoffman Ln.; Chestertown, MD 21620; USA.*
Phone: (410) 778-6868. Fax: (410) 778-9050
sharedearth@aol.com; http://www.sharedearth.org

Shell Corp (Marine Habitat Programme)

The Shell Marine Habitat program is a partnership between the Shell Oil Company and the National Fish and Wildlife Foundation. The purpose of this partnership is to provide grants for projects that benefit marine habitat in and around the Gulf of Mexico.

Contacts: *http://www.nfwf.org/programs/shell/index.cfm*

Shell Foundation

The Foundation focuses efforts on a partnering agreement with the Smithsonian Institution that monitors and assesses the impacts of oil and gas operations on areas of high biological diversity.

Contacts: *http://www.nationalzoo.si.edu*

Sigma X

The Sigma Xi Grants-in-Aid of Research program awards grants of $1,000 or less to undergraduate and graduate students from all areas of the sciences and engineering. Designated funds from the National Academy of Sciences allow for grants of up to $2,500 for astronomy or vision related research. Students use the funding to pay for travel expenses to and from a research site, or for purchase of non-standard laboratory equipment necessary to complete a specific research project. Students from any country are eligible to receive funding. Refer to the website for application procedures.

Contacts: *http://www.sigmaxi.org*

Six Flags Marine World

Six Flags Marine World provides funding of over $10,000 to an average of 1.5 projects per year. 100% of projects are funded ex-situ. Funding requirements are conservation impact and educational. Funding priorities include endangered/threatened species, in collection, endorsed, time frame, single species, in-situ component and husbandry.

Contacts: *http://www.sixflags.com*

Small Grants Programme for Operations to Promote Tropical Forests

The “Small Grants Programme for Operations to Promote Tropical Forests” (SGP PTF) is a European Commission funded initiative implemented by the UNDP and executed by the SEAMEO Regional Centre for Graduate Study and Research in Agriculture (SEARCA). The program is aimed at South-East Asia (and Sri Lanka). The SGP PTF is a community focused, country led programme that works closely with and builds on the existing mechanisms and procedures established by The GEF Small Grants Programme. The main objective of this programme is to encourage and support local communities, who often have little access to financial and technical resources to undertake innovative approaches to sustainable forest management at the local level.

Contacts: *SEARCA College, 4031 Laguna – Philippines*

Phone: +63-49-536.22.90 loc. 395 or loc.418; Fax: +63-49536.24.77

http://www.sgpptf.org

Smith Richardson Foundation

The Smith Richardson Foundation funds research in public policy. The domestic policy program primarily supports research on ways to improve public policies related to welfare reform, economic and regulatory schemes, city governance, and education. The program has a particular interest in research on the effectiveness of public policies and institutions which serve at-risk children and families. The foreign policy program supports projects in national security and defense policy, military history and strategy, the political, military, and economic affairs of the states of the former Soviet Union, economic and security developments in the Asia-Pacific region, and international economic issues of vital interest to American policymakers. There are no application deadlines. Junior Faculty Research Grant Program supports junior faculty engaged in the research and writing of a scholarly book on an issue or topic of interest to the policy community in the areas of American foreign policy, international relations, international security, military policy, and diplomatic and military history. Applicants must have a PhD, hold positions as full-time faculty members, and not have received tenure (or been denied tenure at the university).

Contacts: *Smith Richardson Foundation, 60 Jesup Road, Westport CT 06880, USA*
Phone: (203) 222-6222
http://www.srf.org

SMITHSONIAN INSTITUTION

(i) James e. Webb internship: Annual summer program to promote excellence and diversity in the management of not-for-profit scientific and cultural organizations by placing interns in offices, museums, and research institutes throughout the Smithsonian Institution. Eligible candidates must be graduate and senior undergraduate students who belong to an under-represented minority. Students must be pursuing a degree in Business or Policy Administration. *(ii) Pre-doctoral program awards:* Available to students of conservation, earth sciences, biology, ecology, environmental science, astrophysics, and astronomy. Open to US and non-US citizens, there are a variety of internships and fellowships listed applicable to graduate students. Stipends, requirements and deadlines for awards vary. Refer to website for full details.

Contacts: *http://www.si.edu*

SOCIAL SCIENCE RESEARCH COUNCIL

Work at the SSRC is focused on four program areas: global security and cooperation; Migration; knowledge institutions; and the public sphere. These program areas support working groups, conferences, and fellowships, and a wide range of other research activities. SSRC fellowship and grant programs provide support and professional recognition to innovators within fields, and especially to younger researchers whose work and ideas will have longer-term impact on society and scholarship. These programs often target the spaces between disciplines, where new perspectives emerge and struggle for acceptance, thus ensuring the production of knowledge and expertise on key topics, regions, and social challenges.

Contacts: *Fellowships Office, Social Science Research Council 810 Seventh Ave.; New York City, NY 10019 USA*
Phone: 212 377-2700, Ext. 500. Fax: 212 377-2727
http://www.ssrc.org/fellowships/

SOCIETY FOR CONSERVATION BIOLOGY: SOCIAL SCIENCE WORKING GROUP

The Social Science Working Group (SSWG), with funds provided by the Christiansen Fund, wishes to develop capacity and increase participation in the Society for Conservation Biology (SCB) from conservation social scientists studying and working at African universities or NGOs.

Contacts: *Society for Conservation Biology; 4245 N. Fairfax Drive, Suite 400; Arlington, VA, 22203-1651 USA*
Phone: 703-276-2384; Fax: 703-995-4633
info@conbio.org

SOCIETY FOR THE STUDY OF AMPHIBIANS

The Society lists international funding opportunities for herpetological students and professionals.

Contacts: *http://www.ukans.edu/~ssar/int_grt.html*

SOLON T. KIMBALL AWARD FOR PUBLIC AND APPLIED ANTHROPOLOGY

The Kimball Award offers an opportunity to honor exemplary anthropologists for outstanding recent achievements that have contributed to the development of anthropology as an applied science and have had important impacts on public policy. The anthropological contribution may be theoretical or methodological.

Contacts: *American Anthropological Association; 2200 Wilson Blvd, Suite 600; Arlington, VA 22201 USA.*
Phone 202-785-6597
http://www.aaanet.org

SOPHIE DANFORTH CONSERVATION BIOLOGY FUND

The Sophie Danforth Conservation Biology Fund (SDCBF), established by the Roger Williams Park Zoo and the Rhode Island Zoological Society in 1989, supports conservation programs that protect threatened wildlife and habitats worldwide. Field studies and other projects that demonstrate a multi-disciplinary approach to biodiversity and ecosystem conservation and projects that involve in country collaborators receive the highest funding priority. Environmental education programs, development of techniques that can be used in a natural environment, and captive propagation programs that stress an integrative approach to conservation are also appropriate.

Contacts: *http://www.rogerwilliamsparkzoo.org*

Southern Africa Trust

The Southern Africa Trust supports civil society to engage effectively in regional-level policy processes aimed at overcoming poverty and inequality in southern Africa. The purpose of the Southern African Development Community's (SADC) Regional Indicative Strategic Development Plan (RISDP) is to deepen regional integration in southern Africa. It provides SADC countries with a comprehensive program for harmonizing long-term economic and social policies. It also provides the SADC secretariat and other SADC institutions with a clear view of SADC's priorities.

Contacts: *Southern Africa Trust; PO Box 1234; Halfway House; Midrand 1685 South Africa; http://www.southernafricatrust.org*

Spencer Foundation

The Spencer Foundation supports research that gives promise of yielding new knowledge leading to the improvement of education. The Foundation defines education broadly to include all situations and institutions in which education proceeds, across the entire life-span, in the U.S. and around the world. Three grant programs are available: *(i) major research grants* - provides funding to research projects requiring more than $35,000 in grant support for widely-varied research projects, ranging from medium-sized studies that can be completed in a year by an individual to more extensive collaborative studies that last several years. Applications may be made at and time by submitting an informal letter of inquiry (1,500 words or less) or a brief preliminary proposal. If the proposed research appears to be of potential interest, a more detailed proposal will be requested by the Foundation. *(ii) small grants* - designed to help scholars from a broad range of academic fields at different stages in their careers to build on their past work or to venture into new domains. The program is appropriate for modest-sized research inquiries, exploratory studies, and specific phases of larger investigations, and projects which arise in response to unusual opportunities. Grants generally range from $1,000 to $35,000 and projects may not extend beyond a year in duration. Small grant proposals are welcomed at any time and will be acted upon within approximately three months of receipt. The Foundation's Practitioner Research Communication and Mentoring Program (PRCM) supports research or research capacity-building projects of up to $50,000 over a two-year period. The program seeks to strengthen the effectiveness of teacher/educator-researchers and to clarify the uses of practitioner research. The purpose of PRCM grants is threefold: to establish or strengthen channels for rigorous examination and discussion of the characteristics and knowledge produced by well-constructed teacher/educator-research; to provide teacher/educator-researchers with the opportunity to enhance their research skills through consultation with others in the teacher/educator-research and/or traditional academic communities; and to support high quality teacher/education research projects. Grants made under the PRCM program range from $1,000 to a maximum of $50,000 over a two-year period.

Contacts: *875 North Michigan Avenue, Suite 3930 Chicago, IL 60611-1803*
Phone: (312) 337-7000; Fax: (312) 337-0282
Information@Spencer.org; http://www.spencer.org;

SPILL CONTROL ASSOCIATION OF AMERICA: DR. THOMAS F. DALTON SCHOLARSHIP

The SCAA endowed scholarship is intended to provide financial assistance to students enrolled in university level courses in environmental sciences. In addition to receiving the scholarship, award recipients receive an honorary membership to SCAA and are given the opportunity to interact with organization members. Selections of the scholarship recipients are made primarily based upon academic merit. The number and amount of the scholarship award(s) is made at the discretion of the SCAA.

Contacts: *http://www.scaa-spill.org*

ST. ANDREWS PRIZE FOR THE ENVIRONMENT

The St Andrews Prize for the Environment is an initiative by the University of St Andrews in Scotland and the international integrated energy company, Conoco Phillips. The Prize recognizes significant contributions to environmental conservation, and since its launch in 1998 has attracted entries from more than 50 countries each year on diverse topics including: sustainable development in the Amazon rainforest; urban regeneration; recycling; health and water issues; and renewable energy. Submissions for the annual prize are assessed by a panel of eminent trustees representing science, industry and government with the award going to the project the Trustees consider displays the best combination of good science, economic realism and political acceptability.

Contacts: *The St Andrews Prize Office; University of St Andrews; St. Salvator's, College; St Andrews Fife KY16 9AL United Kingdom,*

Phone: +44 (0) 1334 462 161, Fax: +44 (0) 1334 462 543

prize@standrews.ac.uk

http://www.thestandrewsprize.com

ST. AUGUSTINE ALLIGATOR FARM

The St. Augustine Alligator Farm provides funding of under $1,000 to an average of 2.5 projects per year. 100% of projects are funded ex-situ. Funding priorities include in collection, endorsed and crocodilians.

Contacts: *http://www.alligatorfarm.com*

SUNSET ZOO

Sunset Zoo provides funding of $1,000 to $5,000 to an average of one project per year. 80% of projects are funded in-situ. The project must be endorsed and include conservation impact. Funding priorities include endangered/threatened species, educational, time frame, holistic and in-situ component.

Contacts: *http://www.sunsetzoo.com*

Swedish International Biodiversity Program: Collaborative Programme

The SwedBio Collaborative Programme constitutes the most important avenue for SwedBio to support capacity building in the South and is a key opportunity to directly contribute to development of ideas, methods and policies regarding biodiversity and local livelihood. SwedBio will collaborate with a diversity of strategic initiatives that address three interlinked dimensions: the need to sustainably manage biodiversity and maintain ecosystem services to ensure human well-being and health and contribute to poverty alleviation. the need to ensure equity and human rights in management and use of biodiversity and ecosystem services; the need to support development of appropriate incentive frameworks and good governance in order to address root causes of biodiversity loss.

Contacts: *SwedBio; P.O. Box 7007; SE -750 07 Uppsala, Sweden*
Fax:+46 (0) 18 30 02 46
swedbio@cbm.slu.se
http://www.swedbio.com

Swedish Society for Nature Conservation

SSNC, works on international environmental issues that include disseminating and exchanging information and expertise on the environmental and economic causes of international environmental problems and developing long term co-operation with environmental movements in other countries. SSNC's provides assistance to and co-operates with NGOs through a Small Projects Fund, information exchange, and by engaging in joint campaigns. Current priority is for projects that support the rights of indigenous peoples and local communities to control and manage their natural resources, on the problems caused by industrial tree plantations and the role of Nordic/Swedish forest industry, as well as on sustainable agriculture, genetic resources and food security.

Contacts: *Box 4625, 11691 Stockholm – Sweden*
Phone: +46-8-702.65.00; Fax: +46-8-702.08.55
info@snf.se
http://www.snf.se

Swiss Re: Resource Award

The ReSource Award is an internationally recognized prize for leadership in implementing the principles of sustainability in watershed management. It is conferred annually. The prize money is awarded exclusively for project implementation activities and not for building-up or strengthening the organizations that are submitting an application. The award is aimed at projects which genuinely seek to contribute to raising awareness of the ecological, social and economic significance of water sources and watersheds in developing and emerging countries (only projects that are implemented in a non-OECD country are being considered). Preference is given to those projects which demonstrate innovative preventative measures for protecting water resources, i.e. projects that have a pioneering character in the local context (cultural, institutional or technological), and involve local community and/or regional institutions. Such projects should be based on transferable concepts and thus enable sustained efforts to be made to protect the quality and availability of (drinking) water in other locations.

Contacts: *http://www.swissre.com*

SYSTEM FOR ANALYSIS, RESEARCH AND TRAINING: AFRICAN SMALL GRANTS

In 2003 the International START Secretariat and the Pan African START Committee (PACOM), with funding from the United States National Science Foundation/US Climate Change Science Program (NSF/USCCSP), initiated a program of small research grants to global change scientists in Africa. Awards are made to scientists based at African institutions for research projects related to: *(i)* climate variability and climate change in Africa; *(ii)* impacts/adaptations/vulnerability to global change; *(iii)* land use and ecosystem change; *(iv)* bio-geochemical fluxes, and *(v)* biodiversity. All projects fall within the research frameworks of START's sponsoring programs (IHDP, IGBP, and WCRP). Many of them also contribute to the joint research projects of the earth system science partnership: water systems, the global carbon cycle, and global change and food security.

Contacts: *International START Secretariat, 2000 Florida Ave. NW, Suite 200; Washington, DC 20009 USA*
Phone: 202-462-2213, Fax: 202-457-5859
http://www.start.org

TECH MUSEUM OF INNOVATION: TECH MUSEUM AWARDS

The Tech Museum Awards program honors and awards innovators from around the world who use technology to benefit humanity in the categories of education, equality, economic development, environment, and health. Each year, 25 laureates are honored at a gala dinner, invited to participate in press and media coverage, and introduced to a network of influential advisors. Self nominations are accepted and encouraged. Awards are presented in five categories: health, education, environment, economic development, and equality. The Awards focus on the power of individuals, teams, and organizations around the world and in all sectors. In instances where work is performed in large or well-funded organizations recognition is made to the entrepreneurial individuals and/or teams that fostered innovations which challenged existing paradigms to serve the common good of humanity.

Contacts: *The Tech Museum of Innovation; 201 South Market St. San Jose, CA 95113 USA*
Phone (408) 795-6338
http://www.techawards.org/

THEODORE ROOSEVELT MEMORIAL GRANTS

The Theodore Roosevelt Memorial Grants offer financial support to individuals for research North American fauna in any phase of wildlife conservation or natural history related to the activities of the American Museum. The grant is open to advanced graduate students and postdoctoral researchers.

Contacts: *http://www.research.amnh.org/grants/grantsprog.html*

THIRD WORLD ACADEMY OF SCIENCES

Each year the Academy award prizes to individual scientists from developing countries that have made outstanding contributions to knowledge in the following eight fields of science: biology, chemistry, mathematics, physics, agricultural sciences, earth sciences, engineering sciences, and medical sciences.

Contacts: *TWAS Regional Office for Sub-Saharan Africa (TWAS-ROSSA)*
African Academy of Sciences in Nairobi, Kenya
http://www.nairobi.twas.org

THIRD WORLD ORGANIZATION FOR WOMEN IN SCIENCE

The Third World Organization for Women in Science (TWOWS), with funding from the Department for Research Cooperation (SAREC) of the Swedish International Development Cooperation Agency (SIDA), has instituted a fellowship programme for female students from Sub-Saharan Africa and the least developed countries.The fellowships are for women who wish to pursue postgraduate training leading to a Ph.D. at centres of excellence in the developing countries of the south, other than their own co The fellowships are open to women science graduates, generally below 40 years of age, from countries in Sub-Saharan Africa and least developed countries.

Contacts: *http://www.ictp.trieste.it/~twows/postgrad.html*

TOLEDO ZOOLOGICAL GARDENS

The Toledo Zoological Gardens provides funding of $1000 to $5000 to an average of one project per year. 100% of projects are funded in-situ. The project must include endangered/threatened species, endorsed and conservation impact. Funding priorities include time frame and in-situ component.

Contacts: *http://www.toledozoo.org*

TORONTO ZOO

The Toronto Zoo provides funding of $1000 to $5000 to an average of one project per year. 100% of projects are funded ex-situ. The project must include endangered/threatened species. Funding priorities include in collection, conservation impact, educational, endorsed, time frame, in-situ component, husbandry and public interest potential.

Contacts: *Meadowvale Rd, Toronto, ON M1B 5K7, Canada*
http://www.torontozoo.com

Trees, Water and People

TWP works cooperatively with communities to establish and maintain sustainable forests, watersheds and wetlands while improving people's lives. It coordinates all the projects and provides financial support, technical assistance and training to the organizations in Latin America that are striving to improve the livelihoods of their people and the quality of their environment. TWP's current projects are focused on community reforestation, forest conservation, watershed protection and appropriate technologies, such as fuel-efficient stoves.

Contacts: *633 S. College Avenue, CO 80524, Fort Collins – USA*
Phone: +1-970-484.3678; Fax: +1-970-224.1726
http://www.treeswaterpeople.org

Tropical Ecology Support

TOEB is an organization, which aims to contribute to the ecological sustainability of development projects of the German government. TOEB finances ecological research projects that fall in line with the aims of the GTZ (German Technical Co-operation). It has three program components to co-ordinate (two of which - tropical ecology and tropical forestry - form a single unit and small-scale environment projects).

In Tropical Ecology Research, emphasis is placed on: local resource management (including issues relating to conflict management, land ownership, and land-use rights; economic exploitation of biodiversity (including sustainable tourism and wildlife management); ecological agriculture and agro-biodiversity; utilization of indigenous knowledge; and the use of bio-indicators for monitoring purposes.

Key areas suitable under the tropical forestry component include: instruments and concepts for the rehabilitation and sustainable management of secondary forests and plantations as an alternative to the utilization of primary forests; the tapping, use, and marketing of timber and non-timber resources (e.g. bio-prospecting, tourism, CO2 binding); political, legal, institutional, and social frame-work conditions for forest conservation; dynamic of ecological processes in cases of human intervention in tropical forest systems.

Contacts: *Dag Hammarskjd Weg 1-5, 65760, Eschborn – Germany*
Phone: +49-6196-79.0 Fax: +49-6196-79.1115; http://www.gtz.de/toeb

Tyler Prize for Environmental Achievement

The John and Alice Tyler Prize is awarded for environmental science, energy and medicine conferring great benefit upon mankind. The Prize is endowed by gifts from the Alice C. Tyler Charitable Trust and the John C. Tyler Trust.

Contacts: *Director, The Tyler Prize; University of Southern California; 3616 Trousdale Parkway, AHF 209; Los Angeles, CA 90089-0373 USA*
Phone: 213-740-9760, Fax: 213-740-1313;
http://www.usc.edu/dept/LAS/tylerprize/index.html

UC Pacific Rim Research Programme

The study of the Pacific Rim as a distinctive region. For the purposes of this Program, the term "Pacific Rim" encompasses all areas and nations that border the Pacific Ocean, including all of Southeast Asia and the Pacific islands. Recognizing that the interaction of peoples and states in the region has generated new issues of common concern, the Program places priority on research that is new, specific to the region, and collaborative -- reaching across national boundaries and bridging academic disciplines. Proposals may come from any discipline and should address questions that contribute to an understanding of the Pacific Rim region as a whole. Comparative investigation across national, cultural, linguistic and/or regional boundaries; focus on interactions, flows, or major issues affecting the Pacific Rim region and specific to it; collaboration of scholars in different countries and, where appropriate, different disciplines.

Contacts: *UC Office of the President 1111 Franklin Street, 11th Floor Oakland, CA94607*
Phone: (510) 987-9910; Fax: (510) 987-9456
http://www.ucop.edu

UNEP/CMS Thesis Award on Migratory Species Conservation

The award is offered every three years at the Conference of Parties to CMS at alternating locations. The Secretariat of the Bonn Convention on Migratory Species advertises this award to promote scientific research and conservation of migratory species. The thesis should provide new data and insights into the biology of migratory species or external factors disrupting their migration patterns. Research results must be applicable to conservation measures to the benefit of migratory species.

Contacts: *http://www.cms.int*

UNESCO: MAB Young Scientists Award

The MAB Young Scientists Awards targets young researchers carrying-out interdisciplinary research in line with UNESCO's Man and the Biosphere (MAB) Programme. Priority is given to projects carried-out in biosphere reserves, on the theme of ecosystems and water. The objectives of the award are: to assist exchange of information and experience among a new generation of scientists; to encourage young scientists, in particular those from developing countries, to use UNESCO Man and the Biosphere research, project sites and biosphere reserves in their research; and to encourage young scientists to undertake comparative studies with other sites, in or outside their country. Priority is given to interdisciplinary projects carried out in biosphere reserves on the theme of ecosystems and water.

Contacts: *MAB Secretariat, Division of Ecological and Earth Sciences, UNESCO; 1, rue Miollis; 75732 Paris Cedex 15, France*
http://www.unesco.org/mab

United States Fish and Wildlife Service

(i) This is an agency of the United States government whose goal is to assist countries to become self-sufficient in their capacity to manage their biological resources. To achieve this, the Fish and Wildlife Service supports the Western Hemisphere Program, Asian Elephant Conservation Fund, Small Grants to Russian Federal Nature Reserves, and National Parks. It funds projects that seek to conserve a particular species, education and training in the management of biological resources, community-level environmental education and grass-root activities for the sustained use and conservation of natural resources, technology transfer, and information exchange. *(ii) African elephant conservation fund:* Solicits proposals for the conservation of the African Elephant throughout its range. The Fund supports projects that assist in the conservation by: applied research on African elephant populations and their habitat, including surveys and monitoring; development and execution of elephant conservation management plans; compliance with applicable treaties and laws that prohibit the taking or trade of elephants or regulate the use or management of habitat; conservation education and community outreach; enhanced protection of at-risk elephant populations; efforts to decrease elephant-human conflict; protected area/reserve management in important elephant range; strengthening local capacity to implement conservation plans; trans-frontier conservation plans; wildlife inspection, law enforcement and forensics skills.

Contacts: *4401 North Fairfax Drive ARLSQ 730, VA 22203-1622, Arlington – USA*
Phone: +1-703-358.1754; Fax: +1-703-358.2849
http://www.fws.gov

US National Institutes of Health: International Cooperative Biodiversity Groups

The International Cooperative Biodiversity Groups (ICBG) Program aims to integrate improvement of human health through drug discovery, creation of incentives for conservation of biodiversity, and promotion of scientific research and sustainable economic activity that focuses on environment, health, equity and democracy. This program is based on the belief that discovery and development of pharmaceutical and other useful agents from natural products can, under appropriate circumstances, promote scientific capacity development and economic incentives to conserve the biological resources from which these products are derived.

Contacts: *ICBG Program Officer; Fogarty International Center*

National Institutes of Health, Building 31, B2C39; 31 Center Drive MSC 2220; Bethesda, MD 20892-2220 USA
Phone: (301) 402-9591. Fax: (301) 402-0779

United States-Egypt Joint Science and Technology Fund

Every year the U.S. - Egypt Joint Science and Technology Fund (Joint Fund) sponsors two competitive grant programs: Joint Research Grants and the Junior Scientist Development Visit Grants. These grants provide the opportunity to bring together American and Egyptian scientists to work on a collaborative research project.

Contacts: *American Program Administrator, USAID Building; Ahmed Kamel St. off Laselki St.; New Maadi, Cairo, Egypt*
Phone: +2-02-522-6887; Fax: + 2-02-522-7041; http://www.cairo.usembassy.gov

Utah's Hogle Zoo

Utah's Hogle Zoo provides funding of $1000 to $5000 to an average of two projects per year. 30% of projects are funded in-situ. Funding priorities include endangered/threatened species, in collection, conservation impact, educational, in-situ component and husbandry.

Contacts: *http://www.hoglezoo.org*

Utica Zoo

The Utica Zoo provides funding to an average of three projects per year. 100% of projects are funded ex-situ. Funding priorities include endangered/threatened, conservation impact, educational, time frame, and snow leopard and Mexican wolf.

Contacts: *http://www.uticazoo.org*

Verde Ventures

Verde Ventures is an investment fund managed by Conservation International (CI) that strengthens small and medium-sized enterprises that are strategically important to biodiversity conservation. The fund provides debt and equity financing to select businesses in CI's priority areas. All eligible projects should: make a direct contribution to CI's objectives (outcomes) which include: *(i)* creation of new protected areas; *(ii)* strengthening of existing protected areas; *(iii)* protection of a threatened species and/or; (iv) improvements in biodiversity within a landscape between protected areas.

Contacts: *Verde Ventures; Conservation International; 2011 Crystal Dr., Suite 500; Arlington, VA 22202, USA*
Phone: 703- 341-2400 or 1 (800) 406-2306; Fax: 703-553-0721
verdeventures@conservation.org; http://www.conservation.org

Water Research Commission

The Commission's mandate is to support water research and development as well as the building of a sustainable water research capacity in South Africa. The WRC serves as the country's water-centered knowledge 'hub', leading the creation, dissemination and application of water-centered knowledge, focusing on water resource management, water-linked ecosystems, water use and waste management and water utilization in agriculture. WRC's investment is in building water-centered knowledge.

Contacts: *http://www.wrc.org.za/index.htm*

Wenner Gren Foundation for Anthropological Research

Grants for amounts are available for basic research in all branches of anthropology with particular interest in projects employing comparative perspectives or integrating two or more sub-fields of anthropology. The foundation offers dissertation fieldwork grants, post-PhD grants, as well as the Richard Carley Hunt Postdoctoral Fellowship.

Contacts: *http://www.wennergren.org*

Wetlands International: Wetlands and Poverty Reduction

The Seed Funding Facility provides funds to support conservation and development sectoral organizations to work collaboratively together and engage with local and regional actors in the development of project proposals that address wetlands-poverty issues.

Contacts: *Coordinator; Wetlands and Poverty Reduction Project; Wetlands International P.O. Box 471; 6700 AL, Wageningen Netherlands*
Phone: +31-317-478854 / 8864. Fax: +31-317 8850
http://www.wetlands.org

Wildlife Conservation Society

Since 1895, WCS has worked from its Bronx Zoo headquarters to save wildlife and wild lands throughout the world. WCS is at work across Africa, Asia, Latin America and North America, protecting wild landscapes that are home to a vast variety of species from butterflies to tigers. In 1997 the WCS initiated the Wildlife in Sustainable Forests Program to complement its traditional regional initiatives in Africa, Asia, Latin America, and North America. This program was created to address wildlife conservation in forests managed for timber production.

Contacts: *2300 Southern Boulevard, Bronx, NY 10460, New York – USA*
Phone: +1-718-220.5100; http://www.wcs.org

Wildlife Trust Biodiversity, Ecology and Health Fund

Wildlife Trust's Biodiversity, Ecology and Health Fund (BEHF) supports projects that with a modest level of funding can make an impact on protecting nature or safeguarding ecological health. BEHF projects normally focus on research aimed at improving wildlife conservation science, conservation medicine, natural resources management.

Contacts: *http://www.wildlifetrust.org/ssf.htm*

Woodland Park Zoo

The Woodland Park Zoo provides funding of $5000 to $10000 to an average of eleven projects per year. 90% of projects are funded in-situ. The project must include endangered/threatened species. Funding priorities include in collection, conservation impact, educational, endorsed, time frame, holistic, in-situ component, sponsored by zoo or zoo society staff person, Northwestern U.S. and Pacific Rim.

Contacts: *http://www.zoo.org*

World Parks Endowment

The WPE has played a central role in the creation of new protected areas. It works to conserve endangered rainforests and other biologically rich areas by buying land to create new protected areas and creating endowment funds for tropical park protection and management. The Endowment targets only lands that conserve rare or endangered species and which cost less than $65 an acre.

Contacts: *1616 P Street NW Ste. 200, DC 20036, Washington DC – USA*
Phone: +1-202-939.3808; Fax: +1-202-939.3868
http://www.worldparks.org

World Wildlife Fund Forests for Life Programme Unit

The is the world's largest and most experienced independent conservation organization. Its mission is to protect nature. The Forests for Life Programme Unit underpins WWF's forest conservation work which provides support and leadership in the development and implementation of policy positions and actions at both the global and field level. Of particular concern to WWF are illegal logging and forest crime, conversion of forests to plantation crops of palm oil and soy, forest fires and climate change. WWF's 'Forests for Life' programme is attacking many of these problems with a 3 pronged approach, working to protect, manage and restore the world's forests. This means both on the ground efforts and high level policy work combining to conserve what remains of the world's forests.

Contacts: *Forests for Life Programme, c/o WWF International, Avenue du Mont Blanc, CH-1196 Gland – Swizerland*
http://www.panda.org

THE ZOO

The Zoo provides funding of $1000 to $5000. 95% of projects are funded in-situ. The project must include endangered/threatened, conservation impact, in-situ component and holistic. Funding priorities include single species.

Contacts: *http://www.thezoo.com.au*

Section 4

General Funding Sources

Fastweb Scholarship Search Service

Use this search engine to help find scholarships that pertain to your study of interest.
Contacts: *http://www.fastweb.com*

Foundations

A. Lindsay & Olive B. O'Connor, Inc.
A.L. Mailman Family Foundation, Inc.
Aaron Diamond Foundation, Inc.
Ahmanson Foundation
Albert Kunstadter Family Foundation
Alfred P. Sloan Foundation
Altman Foundation
Amon G. Carter Foundation
Andrew W. Mellon Foundation
Anna Emory Warfield Memorial Fund, Inc.
Annie E. Casey Foundation, Inc.
Arca Foundation
Archie D. and Bertha H. Walker Foundation
Atherton Family Foundation
Axe-Houghton Foundation
Barker Welfare Foundation
Barra Foundation, Inc.
Bay Foundation
Beazley Foundation, Inc.
Beinecke Foundation, Inc.
Ben B. Cheney Foundation
Bingham Trust
Blandin Foundation
Blowitz-Ridgeway Foundation
Boehm Foundation
Booth Ferris Foundation
Bruce L. Crary Foundation, Inc.
Bullitt Foundation, Inc.
Burkitt Foundation
Burnett-Tandy Foundation
Bush Foundation
Bydale Foundation
Camille and Henry Dreyfus Foundation, Inc.
Cannon Foundation, Inc.
Carnegie Corporation of New York
Champlin Foundations
Charles A. Dana Foundation
Charles E. Culpeper Foundation
Charles H. Revson Foundation, Inc.
Charles Stewart Mott Foundation
Charlotte W. Newcombe Foundation
Charlpeg Foundation, Inc.
Christian A. Johnson Endeavor Foundation
Christopher Reynolds Foundation, Inc.
Clarence and Anne Dillon Dunwalke Trust

Claude Worthington Benedum Foundation
Cleveland H. Dodge Foundation, Inc.
Commonwealth Fund
Compton Foundation, Inc.
Cooper Foundation
Cowles Charitable Trust
Daisy Marquis Jones Foundation
Dammann Fund, Inc.
Danforth Foundation
Daphne Seybolt Culpeper Memorial, Inc.
David and Lucile Packard Foundation
Deer Creek Foundation
DeWitt Wallace-Reader's Digest Fund
Dextra Baldwin McGonagle Foundation, Inc.
Dolfinger-McMahon Foundation
Doris and Victor Day Foundation, Inc.
Dorothy U. Dalton Foundation, Inc.
Dorr Foundation
Duke Endowment
Dyer-Ives Foundation
Edna McConnell Clark Foundation
Educational Foundation of America
Edward John Noble Foundation, Inc.
Edward W. Hazen Foundation
El Pomar Foundation
Emil Buehler Foundation
Esther A. & Joseph Klingenstein Fund, Inc.
F. M. Kirby Foundation, Inc.
F.W. Olin Foundation, Inc.
Fan Fox & Leslie R. Samuels Foundation, Inc.
Fannie E. Rippel Foundation
Farny R. Wurlitzer Foundation
Fay J. Lindner Foundation
Florence and John Schumann Foundation
Florence Rogers Charitable Trust
Florence V. Burden Foundation
Ford Foundation
Frear Eleemosynary Trust
Fred Harris Daniels Foundation, Inc.
Fred L. Emerson Foundation, Inc.
Freed Foundation, Inc.
Freedom Forum International, Inc.
Freeman Foundation
Fund for New Jersey
G. N. Wilcox Trust
Gates Foundation
Gebbie Foundation, Inc.
General Service Foundation
George Frederick Jewett Foundation
German Marshall Fund of the U.S.
Gladys Krieble Delmas Foundation
Graham Foundation for Advanced Studies in Fine Arts
Grainger Foundation
Greenburg-May Foundation, Inc.
Greenwall Foundation
Grotto Foundation, Inc.
H. W. Wilson Foundation
Harkness Foundations for Dance
Harold B. & Dorothy A. Snyder, Inc.
Harris Foundation
Harry A. and Margaret D. Towsley Foundation
Harry and Grace Steele Foundation
Hattie M. Strong Foundation
Heckscher Foundation for Children
Helena Rubinstein Foundation
Henry J. Kaiser Family Foundation
Henry Luce Foundation, Inc.
Herbert A. Templeton Foundation
Herman Goldman Foundation
Hillman Foundation, Inc.
Horace W. Goldsmith Foundation
Houston Endowment Inc.
Huber Foundation
Hudson-Webber Foundation
Hugh J. Andersen Foundation
Hyde and Watson Foundation
International Foundation
Ira W. DeCamp Foundation
Ittleson Foundation, Inc.
J. Aron Charitable Foundation, Inc.
J. M. R. Barker Foundation
J.E. and L.E. Mabee Foundation, Inc.
J.M. Kaplan Fund, Inc.
James A. Macdonald Foundation

James S. McDonnell Foundation
Janesville Foundation Inc.
Janet Memorial Foundation
Jerome Foundation
Jessie Smith Noyes Foundation, Inc.
Joe and Emily Lowe Foundation, Inc.
John A. Hartford Foundation, Inc.
John and Mary R. Markle Foundation
John Ben Snow Foundation, Inc.
John D. & Catherine T. MacArthur Foundation
John H. and Wilhelmina D. Harland Charitable Foundation, Inc.
John P. McGovern Foundation
John S. and James L. Knight Foundation
Josiah Macy, Jr. Foundation
Joyce Foundation
Joyce Mertz-Gilmore Foundation
Kresge Foundation
Lannan Foundation
Laura B. Vogler Foundation, Inc.
Lila Wallace-Reader's Digest Fund
Louis and Anne Abrons Foundation, Inc.
Louis and Harold Price Foundation, Inc.
Louis Calder Foundation
Loyola Foundation
Lucille P. Markey Charitable Trust
Lucius N. Littauer Foundation
Lumpkin Foundation
Lynde and Harry Bradley Foundation
Lyndhurst Foundation
M. J. Murdock Charitable Trust
Marmot Foundation
Mary Flagler Cary Charitable Trust
Mary Hillman Jennings Foundation
Mary Livingston Griggs & Mary Griggs Burke
Mary Owen Borden Foundation
Maurice Falk Medical Fund
Max and Victoria Dreyfus Foundation Inc.
McFeely-Rogers Foundation
McGregor Fund
McInerny Foundation
McKnight Foundation
McLean Contributionship
Merck Family Fund
Nathan Cummings Foundation
New York Foundation
Northwest Area Foundation
O.P. and W.E. Edwards Foundation, Inc.
Otto Bremer Foundation
Overbrook Foundation
Patrina Foundation
Paul and Mary Haas Foundation
Peter C. Cornell Trust
Pew Charitable Trusts
Phil Hardin Foundation
Pinkerton Foundation
Plough Foundation
Public Welfare Foundation
Reed Foundation, Inc.
Richard King Mellon Foundation
Richard Lounsbery Foundation, Inc.
Richard Mather Fund
Rittenhouse Foundation
Robert Bowne Foundation, Inc.
Robert Sterling Clark Foundation, Inc.
Robert Wood Johnson Foundation
Rockefeller Brothers Fund
Rockefeller Foundation
Rosenberg Foundation
S. Livingston Mather Charitable Trust
Samuel H. Kress Foundation
Samuel Roberts Noble Foundation, Inc.
Samuel Rubin Foundation
Samuel S. Johnson Foundation
Sasco Foundation
Sequoia Foundation
Shubert Foundation, Inc.
Sid W. Richardson Foundation
Siragusa Foundation
Skillman Foundation
Spencer Foundation
Starr Foundation
Stewart W. & Willma C. Hoyt, Inc.
Stone Foundation, Inc.
Surdna Foundation, Inc.
T.L.L. Temple Foundation
Taconic Foundation, Inc.

Teagle Foundation, Inc.
Town Creek Foundation, Inc.
Trull Foundation
Union Foundation
Uris Brothers Foundation, Inc.
Van Ameringen Foundation, Inc.
Virginia Environmental Endowment
W. K. Kellogg Foundation
W. M. Keck Foundation
Wahlstrom Foundation, Inc.
Wallace Alexander Gerbode Foundation
Walter Scott Foundation
Weingart Foundation
Weyerhaeuser Family Foundation, Inc.
Whitaker Foundation
Whitehead Foundation
William and Flora Hewlett Foundation
William Caspar Graustein Memorial Fund
William E. Scott Foundation
William R. Kenan, Jr. Charitable Trust
William Randolph Hearst Foundation
William T. Grant Foundation
Zellerbach Family Fund.

Please select a Foundation from the list and use the Google search engine (http://www.google.com) to get the relevant contacts, guidelines and eligibility

IUCN - The World Conservation Union

Provides a list of IUCN Corporate Partners in Conservation but does not run any grant programmes.

Contacts: *http://www.iucn.org*

Nike Foundation

Aim to fund projects that enhance the security and empowerment of girls.

Contacts: *http://www.nike.com*

Rotary Foundation

The Foundation offers more than 1,000 grants of variable sizes annually. Check website for additional information on eligibility requirements and deadlines.

Contacts: *One Rotary Center, 1560 Sherman Avenue, Evanston, IL 60201 USA*
Phone: +1(847) 866-3000;
http://www.rotary.org/foundation

W. Alton Jones Foundation

Helps to fund hundreds of environmental groups.

Contacts: *http://www.klamathbasincrisis.org/exposed/w-altonjones.html*

William and Flora Hewlett Foundation

The Foundation concentrates its resources on activities in education, environment, global development, performing arts, and population. The following programs and initiatives are accepting unsolicited letters of inquiry for new grants: Education, Environment, and Performing Arts. See website for program-specific guidelines and to submit a letter of inquiry for a new grant.

Contacts: *The William and Flora Hewlett Foundation; 2121 Sand Hill Road; Menlo Park, CA 94025 USA*
Phone :(650) 234-4500, Fax: (650) 234-4501
http://www.hewlett.org

APPENDICES

APPENDIX 1: RELEVANT CONSERVATION FUNDING SOURCES WEBSITES

http://www.aaanet.org
http://www.aabrahamfoundation.org
http://www.aaionline.org
http://www.aascu.org
http://www.aau.org
http://www.aauw.org
http://www.abcbirds.org
http://www.acbf-pact.org
http://www.acdi-cida.gc.ca
http://www.aciar.gov.au
http://www.acnatsci.org
http://www.acpsec.org
http://www.acs.org
http://www.act.org
http://www.acu.ac.uk
http://www.adf.gov
http://www.aeon.info
http://www.aesf.org/
http://www.af-info.or.jp
http://www.afonet.org
http://www.africaenterprisechallengefund.org
http://www.africanbirdclub.org
http://www.africanconservation.org
http://www.africare.org
http://www.agfund.org
http://www.aibs.org
http://www.aises.org
http://www.ajws.org
http://www.akdn.org
http://www.akglobalfoundation.org
http://www.alcanprizeforsustainability.com
http://www.alcoafoundationfellows.org
http://www.allenfoundation.org
http://www.alligatorfarm.com
http://www.amaraconservation.org
http://www.amazon.com
http://www.ambnairobi.um.dk
http://www.americanantiquarian.org
http://www.amphilsoc.org
http://www.anafeafrica.org
http://www.anchorenv.com
http://www.ancientforests.org
http://www.animalbehavior.org
http://www.animalbehavior.org
http://www.annies.com
http://www.antenna.nl
http://www.aou.org
http://www.asce.org
http://www.asf.ca
http://www.ashdenawards.org
http://www.asp.org
http://www.aspeninst.org

http://www.assets-kenya.org
http://www.astf.net
http://www.aswnet.de
http://www.atlas.org
http://www.atree.org
http://www.aucegypt.edu
http://www.aupha.org
http://www.ausaid.gov.au
http://www.awdf.org
http://www.awf.org
http://www.awra.org
http://www.awu.org
http://www.awwa.org
http://www.aza.org
http://www.baeume.de
http://*www.bamz.org*
http://www.baringfoundation.org.uk
http://www.batbiodiversity.org
http://www.batcon.org
http://www.bbsrc.ac.uk
http://*www.beardsleyzoo.org*
http://www.berkeley.edu
http://www.bgci.org
http://www.binderparkzoo.org
http://www.biodiversityeconomics.org
http://www.biologists.com
http://www.birdlife.org
http://www.birminghamzoo.com
http://www.blacksmithinstitute.org
http://www.bmf.ch
http://www.bmu.de
http://www.bornfree.org.uk
http://www.bou.org
http://www.britishcouncil.org
http://www.britishcouncil.org.pk
http://www.britishecologicalsociety.org
http://www.bronxzoo.com
http://www.brookfieldzoo.org
http://www.brownandcaldwell.com
*http://*www.brownandcaldwell.com
http://www.btinternet.com
http://www.buffalozoo.org
http://www.busac.org
http://www.butler.edu
http://www.cabq.gov
http://www.cairo.usembassy.gov
http://www.canadianembassy.org
http://www.capebiotech.co.za
http://www.capri.cgiar.org
http://www.careforthewild.com
http://www.cargill.com
http://www.carnegie.org
http://www.cbd.int
http://www.cdsintl.org
http://www.cdtfkenya.org
http://www.ceepa.co.za
http://www.cepf.net
http://www.cgiar.org
http://www.chesapeake.org
http://www.chevening.com
http://www.chevron.com
http://www.chevrontexaco.com
http://www.christensenfund.org
http://www.cifor.cgiar.org
http://www.cimonline.de
http://www.cincinnatizoo.org
http://www.cipe.org
http://www.clayoquotbiosphere.org
http://www.clemetzoo.com
http://www.clwr.org
http://www.cms.int
http://www.cnr.berkeley.edu
http://www.co.bergen.nj.us
http://www.col.org
http://www.colobustrust.org
http://www.colszoo.org
http://www.columbia.edu
http://www.commonwealthfoundation.com
http://www.comptonfoundation.org
http://www.conbio.org
http://www.concernamerica.org
http://www.conservation.org
http://www.conservationafrica.com
http://www.conservationfinance.org
http://www.conservationgis.org
http://www.conservationgrants.com
http://www.conservationgrants.com
http://www.conserveafrica.org
http://www.cop.noaa.gov
http://www.cordaid.nl
http://www.cottonwoodfdn.org

http://www.cpb.org
http://www.crawfordfund.org
http://www.croptrust.org
http://www.csc.noaa.govms/fellows.html
http://www.csfund.org
http://www.ctfs.si.edu
http://www.culturalsurvival.org
http://www.damienfoundation.org
http://www.darwin.gov.uk
http://www.davidshepherd.org
http://www.dci.gov.ie
http://www.deepecology.org
http://www.deere.com
http://www.delmas.org
http://www.denverzoo.org
http://www.detroitzoo.org
http://www.developmentmarketplace.org
http://www.dfg.de
http://www.dfid.gov.uk
http://www.dgdc.be
*http://*www.discoveryglobaled.org
http://www.disney.go.com
*http://*www.dm.gov.ae
http://www.dmoz.org
*http://*www.doe.gov
*http://*www.doen.nl
*http://*www.doubleharvest.org
http://www.dow.com
http://www.dreyfus.org
*http://*www.drylands-group.org
http://www.dupontfund.org
*http://*www.duras-project.net
http://www.dwazoo.com
http://www.earthadventure.org
http://www.earthdirectory.net
http://www.earthlink.de
http://www.earthplatform.com
http://www.earthpreservation.org
http://www.earthwatch.org
http://www.eawildlife.org
http://www.echoinggreen.org
http://www.ecoinformatics.org
http://www.ecologic.org
http://www.ecologyandsociety.org
http://www.ecotourismkenya.org
http://www.ecovitality.org
http://www.eeb.uconn.edu
http://www.eirene.org
http://www.elephanttrust.org
http://www.elpnet.org
http://www.elsatrust.org
http://www.en.aswnet.de
*http://*www.endowment.pwcglobal.com
http://www.engr.ku.edu
http://www.environmentprize.com
http://www.epn.org
http://www.EquatorInitiative.org
*http://*www.er.doe.gov
http://www.erefdn.org
http://www.eriezoo.org
http://www.cs.epa.gov
*http://*www.es.epa.gov
http://www.ethnobotany.org
http://www.evolvingearth.org
http://www.ewa.or.at
http://www.explorers.org
http://www.facefoundation.nl
http://www.fanniemaefoundation.org
http://www.fao.org
http://www.farmafrica.org.uk
http://www.farmlandinfo.org
http://www.fas.org
http://www.fastweb.com
http://*www.fauna-flora.org*
http://www.fdncenter.org
http://www.fellowships.aaas.org
http://www.fitzpatrick.uct.ac.za
http://www.floresta.org
http://www.foc-uk.com
http://www.fondationensemble.org
http://www.fordfound.org
http://www.fordifp.net
http://www.forestpeoples.gn.apc.org
http://www.forestsociety.org
http://*www.fossilrim.com*
http://www.fosterscholars.noaa.gov
http://www.foundationcenter.org
http://www.frameweb.org
http://www.fundacionatura.org
http://www.fundeso.org

http://www.futurefoundation.org
http://www.fws.gov
http://www.gaac.org
http://www.gatesfoundation.org
http://www.gbif.org
http://www.gcamerica.org
http://www.gcci.org
http://www.gdnet.org
http://www.gefweb.org
*http://*www.gefweb.org
http://www.geo.de
http://www.geocities.com
http://www.georgewright.org
http://www.gic-gfci.scitech.gc.ca
http://www.global.finland.fi
http://www.global2000.at
http://www.globalfundforwomen.org
http://www.globalgiving.com
http://www.go.worldbank.org
http://www.goldmanprize.org
http://www.gonzalo.er.anl.gov
http://www.gorillafund.org
http://www.gorta.org
http://www.gpz.org
http://www.grantproposal.com
http://www.grants.gov
http://www.grantsmanagement.com
http://www.greengrants.org
http://www.gtz.de
http://www.haribon.org.ph
http://www.healtheffects.org
http://www.heifer.org
http://www.heinz.org
http://www.helvetas.ch
http://www.heritagepreservation.org
http://www.hewlett.org
http://www.hfg.org
http://www.hfp.heinz.org
http://www.hiltonfoundation.org
http://www.hivos.nl
http://www.hoglezoo.org
http://www.honoluluzoo.org
http://www.hosking-tours.co.uk
http://www.hsfi.org
http://www.hudsonriver.org
http://www.iccae.agr.nagoya-u.ac.jp
http://www.icco.nl
http://www.icgeb.trieste.it
http://www.ictp.trieste.it
http://www.icwa.org
http://www.ideawild.org
http://www.idrc.ca
http://www.ifad.org
http://www.ifar4dev.org
http://www.ifaw.org
http://www.ifs.se
http://www.ihdp.uni-bonn.de
http://www.iied.org
http://www.iienetwork.org
http://www.imls.gov
http://www.impumelelo.org.za
http://www.inamori-f.or.jp
http://*www.indianapoliszoo.com*
http://www.inffoundation.org
http://www.intercooperation.ch
http://www.internationalfunders.org
http://www.internationaltreefoundation.org
http://www.irex.org
http://www.isabellafreedman.org
http://www.itilifesciences.com
http://www.itoca.org
http://www.itto.or.jp
http://www.iucea.org
http://www.iucn.org
http://www.iupac.org
http://www.jaf.org
http://www.janegoodall.org.uk
http://www.jca.apc.org
http://www.jec.go.jp
http://www.johnballzoosociety.org
http://www.jpmorganchase.com
http://www.jrsbdf.org
http://www.jsmf.org
http://www.jwrc.or.jp
http://www.kbprize.org
http://www.keidanren.or.jp
http://www.kenyawildlifetrust.org
http://www.kff.org/
http://www.khwarizmi.irost.ir
http://*www.kidszoo.com*
http://www.kijanikenyatrust.org

http://www.kilby.org
http://www.koningsschool.nl
http://www.kuwait-fund.org
http://www.kws.go.ke
http://www.laspau.harvard.edu
http://*www.lazoo.org*
http://www.lcaof.org
http://www.leakeyfoundation.org
http://www.leverhulme.ac.uk
http://www.levinsonfoundation.org
http://www.levistrauss.com
http://www.lewa.org
http://www.lib.msu.edu
http://*www.lincolnzoo.org*
http://www.lindberghfoundation.org
http://www.livingdesert.org
http://www.livingearth.org.uk
http://www.longislandsoundstudy.net
http://*www.lowryparkzoo.com.*
http://www.lpzoo.com
http://www.ludwick.org
*http://*www.macfound.org
http://www.maendeleo-atf.org
*http://*www.maendeleo-atf.org
http://www.mammalsociety.org
http://www.marinebiodiversity.ca
http://www.markle.org
http://www.mazon.org
http://www.mbl.edu
http://www.mcbi.org
http://www.mcf.org
http://www.mcknight.org
http://www.mellon.org
http://www.memphiszoo.org.
http://*www.miamimetrozoo.com*
http://www.mistra.org
http://www.mit.edu
http://www.mkomazi.com
http://*www.mmzoo.org*
http://www.mobot.org
http://www.monsantofund.org
*http://*www.morrisanimalfoundation.org
http://www.mote.org
http://www.mott.org
*http://*www.mountain.org
http://www.ms.dk
http://www.mweka-wildlife.ac.tz
http://www.nabci-us.org
http://www.nafsa.org/content/
http://www.nairobi.daad.de
http://www.nairobi.twas.org
http://www.nationalacademies.org
http://www.nationalgeographic.com
http://www.nationalzoo.si.edu
http://www.nature.nps.gov
http://www.nature.org
http://www.natureandculture.org
http://www.naturekenya.org
http://www.nciucn.nl
http://www.nciv.net
http://*www.nczoo.org*
http://www.nebf.org
http://www.nedcc.org
http://www.neotropicalbirdclub.org
http://www.nepenthes.dk
http://*www.newzoo.org*
http://www.nfwf.org
http://www.nfwf.org
http://www.niehs.nih.gov
http://www.nike.com
http://www.nippon-foundation.or.jp
http://www.nouvelle-planete.ch
http://www.novib.org
http://www.noyes.org
http://www.nrf.ac.za
http://www.nrt-kenya.org
http://www.nsf.gov
http://www.nsgo.seagrant.org
*http://*www.nspe.org
http://www.nspires.nasaprs.com
http://www.nstfawards.org.za
http://www.nt1-ehrweb.aaas.org
http://www.nwf.org
http://www.nwo.nl
http://www.oakfnd.org
http://www.obfs.org
http://*www.okczoo.com*
http://www.on.ec.gc.ca
http://www.opecfund.org
http://www.orau.gov
http://*www.oregonzoo.org*
http://www.oroverde.de

http://www.ots.duke.edu
http://www.overbrook.org
http://www.oxfam.org.uk
*http://*www.pacificenvironment.org
http://www.packard.org
*http://*www.packfound.org
*http://*www.padifoundation.org
*http://*www.paid-wa.org
http://www.palmbeachzoo.org
http://www.panda.org
*http://*www.patagonia.com
*http://*www.pcusa.org
*http://*www.pep-net.org
*http://*www.pepsico.com
http://www.pewoceanscience.org
*http://*www.philipmorrisinternational.com
http://www.phoenixzoo.org
http://www.pin.primate.wisc.edu
http://www.pittsburghzoo.org
*http://*www.plantit2000.com
*http://*www.popcouncil.org
http://www.powerauthority.on.ca
http://www.prem-online.org
http://www.primate.org/
http://www.proform.com.pk
http://www.projectafrica.com
http://www.ptes.org
http://www.publicwelfare.org
http://www.rainforest.org
http://www.rainforest.org.uk
http://www.rainforest-alliance.org
http://www.rainforestconservation.org
http://www.ramsar.org
http://www.ran.org
http://www.rareconservation.org
http://www.rbf.org
http://www.regenwald.org
http://www.regenwaldinstitut.de
http://www.rem.sfu.ca
http://www.research.amnh.org
http://www.research.hq.nasa.gov
http://www.research.hu-berlin.de
http://www.research.uiowa.edu
http://www.research-foundation.org
http://www.rff.org
http://www.rgs.org
http://www.rhinoark.org
http://www.rhinogroup.org.uk
http://www.riverbanks.org
http://www.rockfound.org
http://www.rogerwilliamsparkzoo.org
http://www.rogerwilliamsparkzoo.org
http://www.rolexawards.com
http://www.rosamondgiffordzoo.org
http://www.rotary.org/foundation
http://www.royalcaribbean.com
http://www.royalsoc.ac.uk
http://www.rsnz.org
http://www.rspb.org.uk
http://www.rswr.org
http://www.rufford.org/rsg
http://www.sahel.org.uk
http://www.saheldefs.free.fr
http://www.samuelrubinfoundation.org
http://www.save-the-african-wild-dog.com
http://www.save-the-elephants.org
http://www.savetherhino.org
http://www.savingcranes.org
http://www.sazoo-aq.org
http://www.scaa-spill.org
http://www.scholarships.berkeley.edu
http://www.scholarships.fatomei.com
http://www.scied.science.doe.gov
http://www.scienceinafrica.co.za
http://www.sciencemag.org
http://www.sdc.admin.ch
http://www.sea2.org
http://www.seedinit.org
http://www.seedtree.org
http://www.selfhelpintl.ie
http://www.senecaparkzoo.org
http://www.serconline.org
http://www.setac.org
http://www.sfzoo.org
http://www.sgp.undp.org
http://www.sgpptf.org
http://www.shapingoutcomes.org
http://www.sharedearth.org
http://*www.sheddaquarium.org*
http://www.sheldrickwildlifetrust.org

http://www.si.edu
http://www.sida.org
http://www.siemenpuu.org
http://www.sigmaxi.org
http://www.silcom.com
http://www.snf.se
http://*www.sixflags.com*
http://*www.socioeconomic.org*
http://www.soros.org
http://www.southernafricatrust.org
http://www.spencer.org
http://www.spf.org
http://www.srf.org
http://www.ssrc.org
http://www.start.org
http://www.stetson.edu
http://www.stlzoo.org
http://www.sunsetzoo.com
http://www.surfrider.org
http://www.sustainableharvest.org
http://www.sustainabletravelinternation
al.org
http://www.svenskakyrkan.se
http://www.swbg-conservationfund.org
http://www.swedbio.com
http://www.swissaid.ch
http://www.swissre.com
http://www.tblmirrorfund.com
http://www.techa wards.org
http://www.techsoup.org
http://www.terrehumaine.org
http://www.tgci.com
http://www.thecoca-colacompany.com
http://www.thelawrencefoundation.org
http://www.thestandrewsprize.com
http://www.theworldchallenge.co.uk
http://www.thezoo.com.au
http://www.thresholdtides.org
http://www.toledozoo.org
http://www.torontozoo.com
http://www.toyotafound.or.jp
http://www.trace.org
http://www.treeaid.org.uk
http://www.treefund.org
http://www.treesforlife.org
http://www.treesftf.org
http://www.treeswaterpeople.org
http://www.tropical-biology.org
http://www.tropical-biology.org
http://www.truman.gov
http://www.Ttfkenya.org
http://www.tucsonzoo.org
http://www.turnerfoundation.org
http://www.tusk.org
http://www.ucop.edu
*http://*www.udall.gov
http://www.ukans.edu
http://www.undueinfluence.com
http://www.unep.org
http://www.unep-wcmc.org
http://www.unesco.org
http://www.unisong.com
http://www.universitiesuk.ac.uk
*http://*www.upenn.edu
http://www.upjohninst.org
http://www.usaid.gov
http://www.usc.edu
http://www.users.owt.com
http://www.usip.org
http://www.uticazoo.org
http://www.uwgb.edu
http://www.vi-skogen.com
http://www.vodafonefoundation.org
http://www.volkswagenstiftung.de
http://www.wajones.org
http://www.watamuturtles.com
http://www.waterandfood.org
http://www.wcs.org
http://www.weedenfdn.org
http://www.welthungerhilfe.de
http://www.wennergren.org
http://www.wetlands.org
http://www.wfd.de
http://www.wgf.org
http://www.whitleyaward.org
http://www.wilderness.org
http://www.wildlifedirect.org
http://www.wildlifetrust.org
http://www.wings-globalfund.org
http://www.winrock.org
http://www.wkkfafrica.org
http://www.wn.org

http://www.wolffund.org.il
http://www.woman.ch
http://www.worldbank.org
http://www.worldlandtrust.org
http://www.worldparks.org
http://www.worldvision.org
http://www.wrc.org.za
http://www.wstar.org
http://www.wwf-efn.org
http://www.wwf-uk.org
http://www.zayedprize.org.ae
http://www.zoo.org
http://www.zoofunding.wordpress.com

APPENDIX 2:2008 LIST OF DEVELOPING COUNTRIES

1. Afghanistan
2. Albania
3. Algeria
4. American Samoa
5. Angola
6. Argentina
7. Armenia
8. Azerbaijan
9. Bangladesh
10. Barbados
11. Belarus
12. Belize
13. Benin
14. Bhutan
15. Bolivia
16. Bosnia and Herzegovina
17. Botswana
18. Brazil
19. Bulgaria
20. Burkina Faso
21. Burundi
22. Cambodia
23. Cameroon
24. Cape Verde
25. Central African Republic
26. Chad
27. Chile
28. China
29. Colombia
30. Comoros
31. Congo, Democratic Republic
32. Congo, Republic
33. Costa Rica
34. Côte d'Ivoire
35. Croatia
36. Cuba
37. Chech Republic
38. Djibouti
39. Dominica
40. Dominican Republic
41. Ecuador
42. Egypt, Arab Republic
43. El Salvador
44. Equatorial Guinea
45. Eritrea
46. Estonia
47. Ethiopia
48. Fiji
49. Gabon
50. Gambia
51. Georgia
52. Ghana
53. Grenada
54. Guatemala
55. Guinea
56. Guinea-Bissau
57. Guyana
58. Haiti
59. Honduras
60. Hungary
61. India
62. Indonesia
63. Iran, Islamic Republic
64. Iraq
65. Jamaica
66. Jordan
67. Kazakhstan
68. Kenya

69. Kiribati
70. Korea, Democratic Republic
71. Kyrgyz Republic
72. Lao PDR
73. Latvia
74. Lebanon
75. Lesotho
76. Liberia
77. Libya
78. Lithuania
79. Macedonia, FYR
80. Madagascar
81. Malawi
82. Malaysia
83. Maldives
84. Mali
85. Marshall Islands
86. Mauritania
87. Mauritius
88. Mayotte
89. Mexico
90. Micronesia, Fed. Sts.
91. Moldova
92. Mongolia
93. Morocco
94. Mozambique
95. Myanmar
96. Namibia
97. Nepal
98. Nicaragua
99. Niger
100. Nigeria
101. Northern Mariana Islands
102. Oman
103. Pakistan
104. Palau
105. Panama
106. Papua New Guinea
107. Paraguay
108. Peru
109. Philippines
110. Poland
111. Romania
112. Russian Federation
113. Rwanda
114. Samoa
115. Sao Tome and Principe
116. Senegal
117. Serbia and Montenegro
118. Seychelles
119. Sierra Leone
120. Slovak Republic
121. Solomon Islands
122. Somalia
123. South Africa
124. Sri Lanka
125. St. Kitts and Nevis
126. St. Lucia
127. St. Vincent and the Grenadines
128. Sudan
129. Suriname
130. Swaziland
131. Syrian Arab Republic
132. Tajikistan
133. Tanzania
134. Thailand
135. Timor-Leste
136. Togo
137. Tonga
138. Trinidad and Tobago
139. Tunisia
140. Turkey
141. Turkmenistan
142. Uganda
143. Ukraine
144. Uruguay
145. Uzbekistan
146. Vanuatu
147. Venezuela, RB
148. Vietnam
149. West Bank and Gaza
150. Yemen, Republic
151. Zambia
152. Zimbabwe

Source: World Bank, 7/1/07. Geographic classifications and data reported for geographic regions are for low-income and middle-income economies, as defined by the World Bank. Total of 152 countries.

Index of Funding Organizations

A

B

C

P

R

S

T